Henry Maudslay
& The Pioneers of the Machine Age

Henry Maudslay (1771-1831) by H. Grevedon, 1827. *(Science Museum/SSPL)*

Henry Maudslay
& The Pioneers of the Machine Age

Edited by

John Cantrell & Gillian Cookson

TEMPUS

In Memory of
A.E. Musson
(1920-2001)
*Professor of Economic History
at the University of Manchester
1973-1982*

Front Cover: The Lambeth Works c. 1845. *(Richard Maudslay)*

First published 2002

PUBLISHED IN THE UNITED KINGDOM BY:
Tempus Publishing Ltd
The Mill, Brimscombe Port
Stroud, Gloucestershire GL5 2QG

PUBLISHED IN THE UNITED STATES OF AMERICA BY:
Tempus Publishing Inc.
2 Cumberland Street
Charleston, SC 29401

© John Cantrell & Gillian Cookson, 2002

The right of John Cantrell & Gillian Cookson to be identified as the Authors of this work has been asserted by them in accordance with the Copyrights, Designs and Patents Act 1988.

All rights reserved. No part of this book may be reprinted or reproduced or utilised in any form or by any electronic, mechanical or other means, now known or hereafter invented, including photocopying and recording, or in any information storage or retrieval system, without the permission in writing from the Publishers.

British Library Cataloguing in Publication Data.
A catalogue record for this book is available from the British Library.

ISBN 0 7524 2766 0

Typesetting and origination by Tempus Publishing.
Printed in Great Britain by Midway Colour Print, Wiltshire

Contents

Acknowledgements 6

Notes on Contributors 7

Illustrations 8

Abbreviations 11

1. Introduction *Gillian Cookson* 12

2. Henry Maudslay *John Cantrell* 18

3. The London Engineering Industry at the time of Maudslay
    A.P. Woolrich 39

4. Richard Roberts *Richard L. Hills* 54

5. David Napier *Michael Moss* 74

6. Joseph Clement *A.P. Woolrich* 94

7. Joseph Whitworth *Angus Buchanan* 109

8. James Nasmyth *John Cantrell* 129

9. William Muir *Tim Procter* 147

10. Maudslay, Sons & Field, 1831–1904 *Laurence Ince* 166

Bibliography 185

Index 190

Acknowledgements

The editors and contributors would like to thank the following for their help in the preparation of this book: Christopher Abbott, Tim Ashworth (Local History Librarian, City of Salford), Professor John C. Brown (Astronomer Royal in Scotland), Dr Brenda Buchanan, Alistair Campbell of Airds, Allan Chapman, Michael Chrimes (Head Librarian, Institution of Civil Engineers), Dr Dorothy Clayton, Dr Neil Cookson, Colin Donald, Dr Derek Dow, Penny Feltham (Archivist, The Museum of Science & Industry in Manchester), George Gardner, J. Marc Greuther (Curator of Industry at the Henry Ford Museum), Ann Heath, Prof. John Hume, Prof. Peter Isaac, Mrs Gill Jackson (Library of the Royal Society), Graham Kennison, the Librarian of the Royal Society of Arts, Christopher F. Lindsey, Murdo McDonald (Archivist of the Argyll & Bute Council), Geoff McGarry, Karl Magee (Glasgow District Archive and Library Service), the Technical Librarian and staff of Manchester Central Reference Library, Richard Maudslay, Diana Maudslay, Sean McGibbon, Eleanor Moore (The Museum of Science & Industry in Manchester), Keith Moore (Senior Librarian and Archivist, Institution of Mechanical Engineers), Carol Morgan (Archivist, Institution of Civil Engineers), Virginia Murray (John Murray & Sons), Major Andrew Napier, Colette O'Flaherty (National Library of Ireland), Peter O'Reilly (The British Library Business Information Service), the Hon. Alicia Parsons, Dale Porter (Western Michigan University), Huw Pritchard (Ayrshire Archives), C.J.D. Roberts, the Earl and Countess of Rosse, Iain Russell, Robert Sharp (Archivist at the Science Museum Library), Sara Stevenson (Senior Curator of Photography, Scottish National Portrait Gallery), Emma West, Louise Weymouth (Archivist of Marconi plc), Colin Whitehead, Hilary Williamson (Principal Library Officer, Fine Art Library, Edinburgh Central Library) and Michael Wright (Curator of Mechanical Engineering at the Science Museum).

The Maudslay Society

The editors would also like to thank Dr J. Ian Langford (Secretary of The Maudslay Society) for dealing with a number of enquiries and making information available to us.

The Maudslay Society was founded in 1942 with the primary aim of perpetuating the name of Henry Maudslay. Today the main practical work of the Society is to support a Research Fellowship in Engineering, tenable at Pembroke College, Cambridge. Over the period 1971-2001, nine Maudslay Research Fellows have been appointed. Membership of the Maudslay Society comprises a number of Maudslay Scholars and Fellows, as well as descendants of Henry Maudslay himself and interested students of Maudslay's life and work. Readers can obtain further information from The Maudslay Society, c/o Dr J.C.D. Hickson, Maudslay Scholarship Foundation, Pembroke College, Cambridge.

Notes on the Contributors

Angus Buchanan is Emeritus Professor of the History of Technology at the University of Bath, and Honorary Director of the Centre for the History of Technology. He has recently published a biography of I.K. Brunel. His other research interests include the development of the engineering profession.

John Cantrell is the author of *James Nasmyth and the Bridgewater Foundry* together with a number of articles on the early engineering industry. He is currently editing the Nasmyth correspondence and writing an accompanying biography. He teaches history at The Manchester Grammar School.

Gillian Cookson is editor of the *Victoria County History of Durham* and currently engaged on a history of Darlington. She has written and edited a number of articles and books on nineteenth-century mechanical and electrical engineering, including a biography of the electrical engineer Fleeming Jenkin.

Richard L. Hills was the founding curator of the North Western Museum of Science and Industry in Manchester. He is the author of a number of books on industrial history including *Richard Arkwright* and *Paper Making in Britain, 1488-1988*. He has recently published biographies of Richard Roberts and James Watt.

Laurence Ince has written accounts of the Neath Abbey Iron Company, the South Wales iron industry, the Soho Foundry and the involvement of the Knight family in the early British iron industry. He is head of history at Aston Manor School, Birmingham.

Michael Moss is a research professor in the Faculty of Arts at the University of Glasgow. He has written widely on industrial and business subjects. His books include a history of Standard Life and, with John Hume, a history of Harland & Wolff, the Belfast shipbuilders. He is currently completing a history of the Kennedy family and Culzean castle in Ayrshire.

Tim Procter worked for more than three years as Assistant Archivist at the Institution of Electrical Engineers. He is now working on the Archives of the Soho Project at Birmingham City Archives, where he is cataloguing the business records of Boulton & Watt.

Tony Woolrich trained as a craftsman in the engineering industry and became a professional museum model maker. His publications include studies on industrial espionage in the eighteenth century and technical literature of the eighteenth and early nineteenth centuries. He is series editor of George Watkins' *Stationary Steam Engines of Great Britain*.

List of Illustrations

Fig.
Front Cover　The Lambeth Works
Frontispiece　Henry Maudslay

Chapter Two: Henry Maudslay
1　Henry Maudslay in his study by James Nasmyth, 1830
2　The Maudslay engineers
3　Henry Maudslay's birthplace, Salutation Alley, Woolwich
4　The frontage of Henry Maudslay's Lambeth factory, *c.*1818
5　Henry Maudslay's erecting shop, 1830
6　Maudslay's mortising machine
7　Maudslay's table engine
8　Maudslay's self-tightening collar
9　Maudslay's screw-generating machine
10　Maudslay's bench micrometer
11　Maudslay's screw-cutting lathe, *c.*1797
12　Maudslay's screw-cutting lathe, 1800
13　Maudslay's bench lathe
14　Maudslay's industrial lathe, *c.*1805

Chapter Three: London Engineering at the time of Maudslay
15　Map to show location of the principal engineering firms at the time of Henry Maudslay
16　Joseph Bramah
17　Bramah's hydraulic press
18　John Rennie
19　John Hall
20　Rose engine or figure lathe made by Holtzapffell and Deyerlien
21　Bramah's hydraulic machine for drawing piles
22　Early nineteenth-century workshop drilling machines
23　Early nineteenth-century workshop cylinder-boring machine

Chapter Four: Richard Roberts
24　Richard Roberts
25　Front view of Roberts' gear-cutting machine
26　Roberts' back-geared lathe, *c.*1817
27　Roberts' original planing machine of 1817
28　Roberts' advertisement, 1821
29　Roberts' slotting machine.

Fig.
- 30   A cross-section of part of a self-acting mule headstock
- 31   *Hibernia* sent to the Dublin and Kingston Railway, Ireland, in 1834
- 32   The Atlas Works, *c.*1854
- 33   The Sharp, Roberts 0-4-2
- 34   Roberts' small 'portable' punching and shearing machine
- 35   The Jacquard plate-punching machine, 1847

Chapter Five: David Napier
- 36   David Napier in his old age
- 37   An abbreviated Napier family tree
- 38   Napier's drawing of a press manufactured by Maudslay
- 39   Napier's tracing machine drawn by Joseph Clement
- 40   The *Nay-Peer* press
- 41   A power-driven perfecting printing machine, *c.*1828
- 42   Napier's hand-cranked bullet-making machine
- 43   The hydraulically driven *Automaton* coin-weighing and classifying machines
- 44   James Murdoch Napier
- 45   An ornately mounted Captain's Registering Compass

Chapter Six: Joseph Clement
- 46   Clement's instrument for drawing ellipses
- 47   An example of the use of Clement's instrument for drawing ellipses
- 48   Clement's drawing table for large mechanical and architectural drawings
- 49   Clement's turning lathe
- 50   Clement's planing machine (side elevation)
- 51   Clement's planing machine (end elevation)
- 52   Clement's driver
- 53   Babbage's difference engine

Chapter Seven: Joseph Whitworth
- 54   Sir Joseph Whitworth
- 55   Whitworth's street cleaning machine
- 56   Whitworth's cylindrical gauges
- 57   Whitworth's measuring machine
- 58   Whitworth's hand screwing apparatus
- 59   Whitworth's patent railway wheel turning lathe
- 60   Whitworth's self-acting planing machine
- 61   Whitworth's bolt- and nut-screwing machine
- 62   Whitworth's self-acting radial drilling and boring machine
- 63   Whitworth's self-acting compound slotting machine
- 64   Whitworth's universal shaping machine
- 65   Whitworth letter and attached portrait

Fig.

Chapter Eight: James Nasmyth

66 James Nasmyth by D.O. Hill, *c.*1845
67 Nasmyth's road steam carriage
68 Nasmyth's nut-cutting machine
69 Nasmyth's Dale Street workshop
70 The Bridgewater Foundry
71 Nasmyth's shaping machine
72 Nasmyth's steam hammer engine
73 Nasmyth's Scheme Book sketch of the steam hammer
74 Nasmyth's steam hammer
75 The traditional and safety foundry ladle

Chapter Nine: William Muir

76 William Muir
77 Muir's advertisement of 1850
78 The Britannia Works, *c.*1895
79 Muir's patent 7in centre double-geared lathe
80 Muir's self-acting radial drilling machine
81 Muir's self-acting, surfacing and screw-cutting lathe, *c.*1895
82 W. Muir & Co. copying presses with Muir's patent stands, *c.*1890
83 Muir's patent double grindstone apparatus, *c.*1895
84 Muir's machine for shaping barrel bedding, *c.*1890

Chapter Ten: Maudslay, Sons & Field, 1831-1904

85 Joshua Field
86 Joseph Maudslay
87 The Maudslay paddle engines for *HMS Retribution*, 1842
88 The horizontal double piston rod engine fitted by Maudslay's into the *Jumna*
89 A 16hp six-column beam engine built by Maudslay's in the middle of the nineteenth century
90 Details of a large table engine, 1841
91 Charles Sells
92 Details of a Maudslay Cornish beam pumping engine
93 Walter Henry Maudslay.
94 Maudslay's triple expansion engines, 1895
95 The Maudslay's Great Wheel at Earl's Court, 1895

Abbreviations

Libraries and record offices

BCA	Birmingham City Archives
BL	British Library
ECA	Edinburgh City Archives
GUA	Glasgow University Archives
JRUL	John Rylands University Library, Manchester
LMA	London Metropolitan Archives
MPL	Manchester Public Libraries
MSIM	Museum of Science and Industry, Manchester
NAS	National Archives of Scotland
NLS	National Library of Scotland
PRO	Public Record Office
ROE	Royal Observatory, Edinburgh
SA	Salford Archives
SMAC	Science Museum, Archive Collection
SSPL	Science & Society Picture Library, Science Museum
UBL	University of Bristol Library

Published sources

DNB	*Dictionary of National Biography*
PICE	*Minutes of Proceedings of the Institution of Civil Engineers*
PIME	*Minutes of Proceedings of the Institution of Mechanical Engineers*
TNS	*Transactions of the Newcomen Society*
TSA	*Transactions of the Society of Arts*

1
Introduction
Gillian Cookson

This book took shape from an initial idea to investigate the network of engineers working in Manchester at the time of Whitworth, Nasmyth and Roberts. It was clear that Manchester's vibrant industry acted as a magnet for any ambitious mechanical engineer of that era. Yet as we began to examine the connections between some of the leading participants, all roads led back to London, and more specifically to the shop of Henry Maudslay.

So the project turned around, to start at the centre of Maudslay's rippling influence. The result is a collection of biographies of Maudslay's most prominent associates which describes the careers of pioneers in machine tool technology, and also traces the assimilation of Maudslay's ideas and methods.[1] To provide the context within which this innovative and influential generation of engineers worked, the opening chapters describe the career of the man himself, Henry Maudslay, and give an account of the London engineering industry in his time. The book concludes with a chapter following the history of Maudslay's business after his death.

Contributors to this volume present previously unpublished documentary research, as well as original analysis which has helped clarify, for instance, the genesis of certain machine tools in an age when innovation was a matter of claim and counter claim. It is still possible to examine many of these machines in the Science Museum and other collections, and there are respectable histories of their technical development, notably Rolt's *Tools for the Job*, first published in 1965. Information about the circumstances in which the machines were designed and built is, however, more difficult to establish. Technological innovation proceeds incrementally, and simultaneous 'invention' or discovery, where individuals working independently produce the same concept at the same time, is quite common. Many developments have been wrongly attributed, or full credit given to someone only partly responsible for an innovation. The reality is seldom clear-cut, and often elusive, although careful research can expose the way in which myths have growth from half truths, or from a careless reading of Samuel Smiles.

Smiles in particular casts a long shadow across nineteenth-century engineering. While he was responsible for recording valuable detail which would otherwise have been lost, Smiles was essentially a caricaturist. Nor can anyone so motivated by social and political ends be seen as a dispassionate observer. His strong beliefs clouded his judgment of the men themselves and of their achievements. For instance, as editor of Nasmyth's autobiography Smiles failed to inject any objectivity into the account of the man's achievements. As a journalist, he was above all interested in stories of 'triumph over adversity', of single-mindedness, dedication and the rise from humble origins. He set an agenda which still tends to define those engineers that are most worthy of discussion. So while on the one hand Smiles has supplied invaluable raw material for some of the chapters in this book, he is also responsible for the fact that other engineers have been marginalised and their contributions overlooked. By implication, anyone

Introduction

excluded from the Smilesian 'communion of engineering saints' does not merit further analysis.[2]

The negative influence of Smiles may therefore be responsible for the fact that David Napier and William Muir have received only brief notice from later writers.[3] On the other hand, some of those selected for the Smiles treatment have had little subsequent attention. This could reflect Smiles' unevenness of treatment, for example in giving only a few pages to engineers as prominent as Whitworth and Roberts, while Nasmyth and Maudslay, and even Joseph Clement, were considered to warrant complete chapters in *Industrial Biography*. Yet Clement has not subsequently been considered by biographers. Even Henry Maudslay himself has not attracted a full-length biography in English.[4] The sole full-scale account of Whitworth is not of scholarly quality, so that much of the apparently new information it presents is impossible to verify.[5] Only James Nasmyth, whose autobiography has been followed by thorough accounts of his life and business by John Cantrell, and Richard Roberts, with a number of modern articles assessing his career and technical achievements, and now the subject of a new full-length study by Richard Hills, have received anything near their biographical dues.

The shortcomings of Smiles are only part of the explanation. Some of the engineers themselves and their associates were responsible for playing down the contribution made by colleagues: Whitworth was written out by Nasmyth, Clement by Babbage. There is also a fairly general shortage of business archives before 1850, and engineering is particularly badly represented among surviving records. Of the subjects in this volume, only Nasmyth left a substantial archive, and that is dispersed between Salford Archives, the British Library, the National Library of Scotland and the Public Record Office. For Maudslay there are a few papers in the Science Museum library. Little or nothing survives to document the businesses of the remainder. This is not to say, of course, that there is no source material. Some of the chapters in this collection illustrate just how much it is possible to reconstruct industrial history from obituaries, patent records, newspapers, directories, technical journals, census enumerators' lists, parliamentary papers and a variety of other biographical sources.[6] There remains, though, a tantalising gap in the information available about day-to-day working in engineering, about how technical knowledge was regarded, how it was disseminated and how it advanced. One possibility of an assemblage of career histories such as this – which inform us about individuals' personalities, particular talents, motivations and other interests – is that it may help define the relationships at the heart of early mechanical engineering. Some sources address this issue directly, such as Nasmyth's notes on 'Maudslay's Maxims', his methods of working and attitude towards his workforce. Other evidence, at times here painstakingly assembled, offers the historian a basis for speculation about the driving forces behind industrial innovation.

The group which has been called the Maudslay nursery[7] was far from homogenous. In fact the engineers identified here as Maudslay's most eminent graduates fit into two, chronologically separated, groups. Roberts was with Maudslay from 1814 to 1816; Napier from approximately 1814 to 1815; Clement from about 1815 to 1817. Whitworth was there around 1826, Nasmyth from 1829 until after Henry Maudslay's death, and Muir did not arrive until two months after Maudslay had died. Their origins were similarly diverse. Roberts and Clement both had amateurish backgrounds in wood- and metal-working, although both could draw, then a rare and prized skill in a mechanical engineer. Napier, their contemporary at the London firm, was from a long-established and numerous engineering family and served a conventional apprenticeship, although even in an engineering dynasty, such a formal training cannot be assumed – Napier's cousin, a man of comparable stature, had never been indentured. Of the later cohort, only Muir seems to have had a formal apprenticeship, though this was for only five-and-a-half rather than the traditional seven years. Whitworth moved between employers at the age of about eighteen, whereas apprenticeships normally started at fourteen and ended when the boy became a man at twenty-one. Nasmyth, the most middle-class of the Maudslay school, arrived at the London shop as special assistant to Maudslay – more a pupil than an apprentice, although not formally known as such. He had learnt his skills at home and, coming from a family of artists, had acquired exceptional proficiency in draughtsmanship. By the late 1820s, there had developed some formality in the training of shop-floor mechanical engineers; the preceding generation of machine-makers had had little more than a general training as smiths or founders, combined with exposure to the rapid changes underway in both products and techniques. But the range of experience in the young men arriving at Lambeth from the provinces, reflects a continuing dynamism in the industry. It is notable too how many of Maudslay's best men, despite the disparity in their backgrounds and the rather random ways in which they had acquired the skill, were talented draughtsmen.

But why did this motley collection of young provincial English and Scottish mechanics gather at Maudslay's? The general answer to this is that improvers, journeymen who had recently completed their initial training in engineering, would follow an itinerant life as a means of broadening their skills and technical knowledge. Boys who had worked only in small workshops were ambitious to experience life in the most modern and well-equipped factories, as a part of their post-apprenticeship education.[8] Millwrights had a longer tradition of frequent travels at home and abroad, partly because their work was site-based, but also as a means of extending their skills.[9] Maybe the practice spread to the newer branches of mechanical engineering. Certainly Clement and Roberts were much travelled before arriving at Maudslay's. Napier, despite the deep engineering roots of his family, and the range of activities which their businesses encompassed, had also moved around extensively. Michael Moss suggests this may have been a means of acquiring information, to which Maudslay did not object as he was well-disposed towards the Napier family. But perhaps too much has been made of the idea of industrial espionage in early engineering. Spying may be an anachronistic concept here, an unduly sinister explanation for an innocent and necessary process of learning. The habit in mechanical engineering, where knowledge advanced far more quickly than any written means of disseminating it, was for young journeymen to learn through their travels. The London engineering chapter illustrates the advantages which the capital enjoyed as a centre of engineering – especially the large local demand for engineering products, both military and civil. Moreover, it shows the role of London as a magnet for engineers from the provinces and continent, a role which continued long after other centres of engineering had developed in provincial industrial towns.

The specific attraction of Maudslay's shop was, of course, its reputation. Richard Roberts summarised its tangible benefits: learning how best to use tools, acquiring skills by mixing with the best workmen, and soaking up the 'spirit of active contrivance' which was pervasive.[10] The 'nursery' benefitted men at different stages in their careers, some employed as various grades of trainee, some journeymen, and some other types of associate. Isambard Kingdom Brunel claimed to have gained 'all my early knowledge of mechanics' from Maudslay,[11] evidently treating his time there as a subsidiary education for one who planned to enter a different branch of engineering, at a time when universities did not offer courses in the subject.

What were the benefits of all this to Maudslay? The waves of itinerant workers may, at the very least, have helped his firm maintain a level of production at a time when engineering skills were in extremely short supply. He also learnt from them, identified their talents and used them to develop aspects of his production. Whitworth helped solve a planing problem and Clement is supposed to have trained Maudslay's drawing office in improved techniques. Later, when his fledglings had flown the Lambeth nest and become established on their own, Maudslay had the advantage of acquiring machine tools from them, and would presumably have been very well-informed on the quality of what he was buying. Whitworth and Roberts machines were recorded as working in Lambeth alongside Maudslay's own.[12]

Personal relationships between the Maudslay engineers are in most cases a matter for conjecture.[13] Maudslay remained on friendly terms with Clement, and was eulogized by Nasmyth. Nasmyth wrote favourably of Roberts, Whitworth employed Muir for a time and Clement produced drawings for Napier.[14] A number of the engineers gravitated to Manchester, for reasons discussed by Angus Buchanan in his chapter on Whitworth[15] – although proximity did not necessarily equate to closeness. Nasmyth and Whitworth appear to have disliked each other, though the evidence for this is entirely circumstantial, and based on the idea that they pointedly ignored each other's achievements. There is no positive confirmation of their antipathy. The engineers' other practices and beliefs were as varied as their social and technical backgrounds. Whitworth and Muir promoted share schemes among their workers, contrasting with the more robust management style of Nasmyth. Though most were very innovative, their use of the patent system differed. Roberts took out twenty-five, while Maudslay himself had held only six, and Clement none at all. As for their mature views on the education of engineers, while Whitworth became synonymous in Manchester with the development of college-based technical education, Nasmyth held stubbornly to the idea of 'on the job' training: 'I have no faith in Royal 'Kademys' of Music nor Colleges of Technical Education with suitered Professors &c &c &c[.] Workshops are the true and only Colleges for such practical Education'.[16]

Perhaps the greatest contrast between these men is manifested in their business aptitude and commercial fortunes. While all were talented engineers, some had apparently little interest in carving out great wealth, while others were incapable of holding on to the capital they had made. Roberts was the prime example of commercial ineptitude. Although at times in his career he made large amounts of money, it was all squandered ultimately in his pursuit of ill-advised innovation. Engineering took precedence over financial security. It would be too blunt to say that creative ability in design or production was enjoyed at a cost of profit – the choice was not clear-cut, and some of the engineers managed the balance admirably. Some made a great commercial success in the development of mass production systems and inter-changeable parts, by constructing the machines and systems which would deliver this: Clement, Maudslay, Whitworth – to be associated forever with best practice, standards and measurement – and Nasmyth, whose relatively short career made him the contemporary equivalent of a multimillionaire from the steam hammer and locomotives, a fortune which he doubled by post-retirement speculations. Others did well by using mass-production to make other machines, for example Roberts during the phase of his career when he successfully mass-produced looms.

Yet constant innovation was evidently a mistake in business terms. It explains the discrepancy between the mechanical talent of Roberts and the long-term financial rewards he enjoyed. Clement, another noted innovator, enjoyed only modest commercial success, despite a continuing concern to exact his just rewards from Babbage and other associates. The probate valuation of his estate was about £12,000, which may be considered a rare achievement for a weaver's son who had not even served a recognised apprenticeship,[17] but in the scale of profits made by some of his contemporaries, and in the context of what we now know was possible for engineers of his background and talent, it was a relatively small success. The unsung Muir made high quality products with innovative features, and continued experimentation into his old age. He drew a line, however, at risking his business in the quest for mechanical novelty, and the firm he established was solid enough to survive half a century after his death. Specialisation did the trick for the long-term future of Maudslay's firm, as it was able to continue for decades on the strength of its expertise in steam, especially marine, engines and boilers. Yet while Napier too specialised, making printing and related machinery, and although his son was said to have had a greater mechanical talent still, his firm did not achieve the financial results which may have been expected, because father and son both lacked a basic understanding of accounts. It is significant that Nasmyth's period of innovation was short and to the point. Experimentation stopped once the steam hammer was in commercial production, and Nasmyth's product range of machine tools and steam engines remained substantially unchanged between 1839 and his retirement in 1856.

As the machine tool technology upon which Maudslay left his indelible mark moves on, so too does the study of history. Most previous accounts of the early toolmakers, by the likes of Rolt, Steeds, Dickinson and Gilbert, have been driven by a desire to appreciate and describe the evolving technology. A.E. Musson, who died in October 2001 while this volume was in progress, was arguably the first academic historian to examine in any detail the contribution of Nasmyth, Whitworth, Roberts, and others, from the perspective of their role in a wider industrialising world.[18] In other words he moved away from a mainly technological standpoint, to consider the early machine tool makers in their industrial context. The approach we have adopted will, we hope, be the means of deepening understanding further still, using the wide lens of biography to explore the technological, business and personal progress of the men who made the machines which made the industrial revolution.

Endnotes

[1] Not all of Maudslay's connections can sustain a full chapter to themselves. Samuel Seaward is discussed in the chapter on London engineering; for Francis Lewis and George Nasmyth, see Cantrell 2002.
[2] See Jarvis 1997.
[3] There has never been a separate study of Napier, other than the assessment in Wilson and Reader 1958; the only previous study of Muir is the memoir by Robert Smiles.
[4] There is a biography in Russian: Zagorskii and Zagorskaia 1981. Details of a CD ROM published in

2001 by the Kew Bridge Steam Museum, which includes a collection of papers on aspects of Maudslay's life and work, are given in the bibliography, below.

[5] Atkinson 1996.
[6] For more on this, see Cookson 1998, pp.27-35.
[7] The phrase is Armytage's: 1961, p.127.
[8] See for instance the autobiography of Thomas Wood of Bingley, who was ambitious to work for Hibbert and Platt of Oldham: Burnett (ed.) 1974, pp.310-12.
[9] A prominent instance is Peter Fairbairn: see Cookson 1994, pp.129-31, for discussion of this and for further examples.
[10] See p.55 below.
[11] I.K. Brunel to Messrs Maudslay and Field, 24 March 1853: quoted in Buchanan 2002, p.236.
[12] Wilson and Reader 1958, p.26.
[13] This issue is explored in more detail in Cantrell 2001.
[14] See Fig.39.
[15] See p.112 below.
[16] J. Nasmyth to S. Smiles, 31 March 1882: BL, Add. Mss., 71076 f.116.
[17] PRO, IR26/1668 f.248.
[18] Notably with Eric Robinson, 1969; see also Musson 1972 and 1980.

2
Henry Maudslay
John Cantrell

By the time that Henry Maudslay died in February 1831 he was widely regarded as one of the greatest engineers of his generation.[1] Indeed William Fairbairn claimed that Maudslay was one of six engineers who completely dominated the profession between 1790 and 1830 to the extent that 'scarcely any work of importance was accomplished without one or other of them having been consulted'.[2] A Prussian publication of 1833 described Maudslay as 'one of the best mechanics in England'.[3] This reputation was based partly on the fact that as a master craftsman Maudslay appears to have had no equal; his talents in this direction were recognised by other leading engineers such as Joseph Bramah and Marc Brunel when they engaged him to give practical form to their mechanical ideas. It was also founded on his inventive skills – both the self-tightening collar for the hydraulic press and the table engine would, on their own, have gained him a significant place in the engineering annals. But perhaps he is best known for the way in which he promoted good workshop practice, insisting on the deployment of slide rests, standardising screw threads, appreciating the importance of true plane surfaces and precise measurement. These new standards and objectives were disseminated and improved by Maudslay's disciples so that the ultimate memorial to his work lay in the work of Whitworth, Roberts, Nasmyth and other engineers noted in this book. As they in turn passed on the new practices to their own apprentices and workmen, a significant portion of Britain's engineering heritage in the nineteenth and twentieth centuries can be traced back to Maudslay's teaching and example.

As with many of his students such as Roberts and Whitworth, Maudslay's origins were relatively humble. In the sixteenth, seventeenth and eighteenth centuries the Maudsley family was based around Clapham in North Yorkshire, with many of the male members described as yeoman or husbandman. Maudslay's father, also named Henry, was born in 1725. As a young man he had got into some trouble in reference to an illegitimate child 'sworn to him' and in consequence left the district and made for Norwich where he enlisted as a wheelwright in the Royal Artillery.[4] From this occupation he was invalided out following a musket ball wound to his neck in the West Indies. Next he found employment as a storekeeper at the Woolwich Dockyard. Henry Maudsley[5] senior married Margaret Laundy (1737-1792), a widow and the daughter of William Whitaker of Ringwood. This union produced seven children and Henry Maudslay, the subject of this study, was born on 22 August 1771.[6]

Henry Maudslay junior obtained his first employment at the age of twelve as a powder monkey, filling cartridges at the Woolwich Arsenal. Two years later he was transferred to the carpenter's shop but showed much more interest in the activities of the nearby smithy. Accordingly he was moved to the smith's department at around the age of fifteen. From this point onwards Maudslay's life was dedicated to working with metal. During the course of the next three years Maudslay acquired such skill in forging and metal working that he was singled out by Joseph Bramah who was looking for a craftsman capable of manufacturing his

Fig.1 Henry Maudslay in his study by James Nasmyth, 1830. *(The Fine Art Library, Edinburgh City Libraries and Information Services)*

patent lock. Maudslay remained with Bramah between approximately 1790 and 1797, later becoming foreman of the works, and married Bramah's housekeeper, Sarah Tindale (1762-1828). The marriage resulted in four children before the end of 1797 and, requiring an increase in his wages of 30s per week to support this growing family, Maudslay applied to his master for a rise. When this was refused he decided to set up business on his own and rented a workshop and smithy at 64 Wells Street, off Oxford Street.

His first customer was an artist who gave him an order for some iron work with a new mechanism for a large easel. Other orders soon poured in for 'he had a wide known fame even then as a first rate workman' and a note of 1799 refers to 'Mr Maudsleys the tool makers'.[7] Larger premises soon became a necessity and in 1802 he moved to 78 Margaret Street, Cavendish Square.[8] The business expanded so that he was able to employ eighty men[9] before making his final move to Lambeth Marsh in 1810 with a frontage on Westminster Bridge Road. At this point Maudslay was the sole proprietor of the business but by 1812 he was in partnership with John Mendham, the uncle of Matilda Field, Joshua Field's wife, as indicated by a capital balance sheet.[10] This shows that in 1812 the business comprising 'Premises Machinery &c.' was valued at £31,500 of which John Mendham's share was one third or £12,500. In that year total profits amounted to £4,500 which was allocated 2:1 in favour of Maudslay. The same account sheet shows that profits increased to £6,798 in 1813 but then decreased to £5,520 in 1814. It was in this year that J.G. May, a Prussian Factory Commissioner recorded that Maudslay 'constructs a great variety of machines' and 'employs over 200 workers'.[11] In 1815, a year of recession, profits slumped to £2,100. Thomas Guppy chose this time to present himself to Maudslay in the hope of becoming an apprentice. The

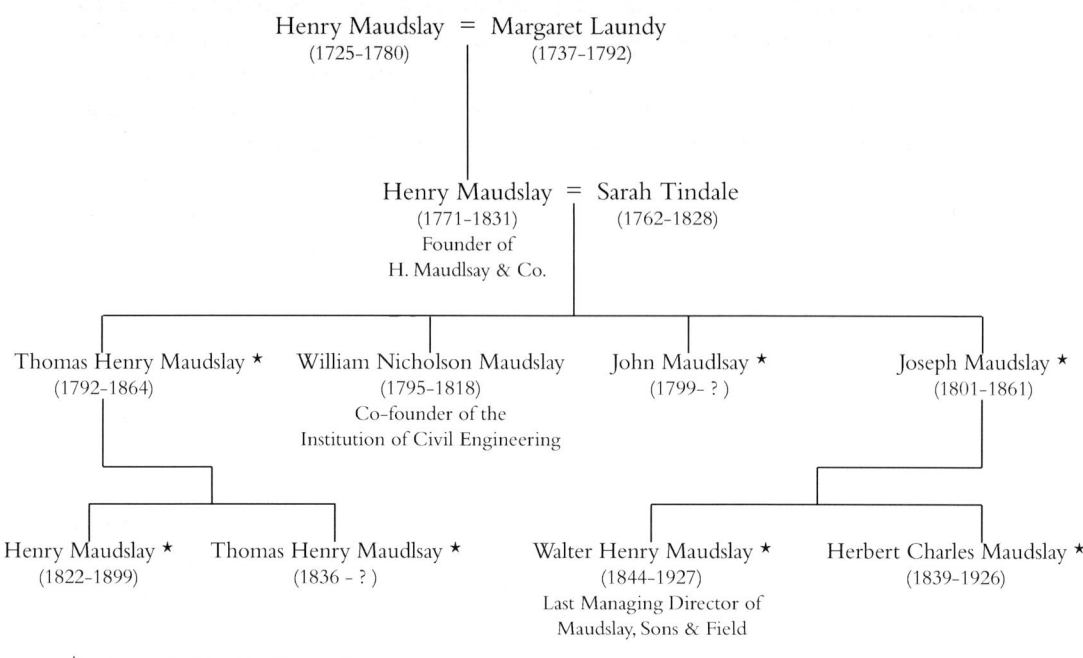

Fig.2 The Maudslay Engineers.

latter, pointing to the deserted workshops and rusty tools said 'The war is at an end; the Government no longer gives me work; all my earnings have been spent in making tools, and in the extension of my works, and now I am about to be ruined'.[12] This was to be but a very temporary setback, for profits in 1816 increased more than threefold to £6,938. There is no mention on the account sheet of any other partners though it has been suggested that Maudslay was also in partnership with his eldest son, Thomas Henry Maudslay, together with Joshua Field, from as early as 1812.[13] Certainly at about that time the firm became known as Henry Maudslay & Co.[14] but there is no definite evidence for the inclusion of others until December 1820, when a notice of partnership dissolution between Maudslay, his son, Field and Mendham appeared in the *London Gazette*.[15] What is not known is when that partnership was formed. The account sheet suggests that this did not happen until 1818 at the earliest – since before that date all the profits were distributed between Henry Maudslay and Mendham. Mendham left the partnership in 1820 and for most of the rest of Maudslay's life the firm traded as Maudslay, Son & Field. Codicils to Maudslay's will make it clear that some time between 10 January and 3 February 1831 Maudslay's sons, John and Joseph, were admitted to the partnership and the firm became Maudslay, Sons & Field.[16] During the 1820s the firm clearly prospered and one contemporary account referred to 'the extensive factory of Messrs Maudslays, supposed to be the most complete in the kingdom. Steam engines, tanks for shipping, and all works connected with various factories, are here executed in the best manner. They occasionally employ upward of two hundred men'.[17] Maudslay's business was clearly that of general engineers and among the orders accepted by the firm before the founder's death were those for marine engines, stationary steam engines, saw mill machinery, gun-boring machinery, coin-minting machinery, hydraulic presses, bronze foundry equipment for the sculptor, Chantrey, and the shield for the Thames Tunnel.

The foundations for Maudslay's reputation as a master craftsman were laid in the carpenter's shop and smithy of the Woolwich Arsenal. Within four years Maudslay progressed from the status of novice to that of one of the most sought after mechanics in London, despite the fact that he never served a formal apprenticeship. At the Arsenal smithy he learnt to work with hammer, anvil, file and chisel and became especially skilful in the forging of light iron work. A favourite job of his was that of forging trivets 'all out of the solid' from Spanish iron bolts. This earned him the surprised admiration of the old experienced hands and caused his reputation to spread to other smithies in London. At about this time Joseph Bramah was experiencing a difficulty in satisfactorily mechanising the manufacture of his famous lock, patented in 1784. Bramah consulted an employee of William Moodie, an 'old German workman...who was esteemed the most ingenious man in the trade of his day,' but without success. It was at this point that another of Moodie's workmen recommended Maudslay. That Maudslay was able to resolve Bramah's problem and earn himself an offer of employment in 'the only place

Fig.3 Henry Maudslay's birthplace, Salutation Alley, Woolwich. *(Vincent, 1888-1890)*

Fig.4 The frontage of Henry Maudslay's Lambeth factory, c.1818. *(Lambeth Archives)*

then in London where any thing like engineering mechanism was carried on' was a clear sign of his exceptional mechanical talent. He also had considerable confidence in his own abilities, for when Bramah first gave him employment he was clearly worried that this 'slender looking lad' with an absence of formal training would be unable to gain the respect of the established workmen. Maudslay resolved this problem too by pointing to an old worn out bench vice and offering to recondition it within the day. The offer was accepted and after a programme of re-steeling, filing, cutting, hardenening, tempering, cleaning and resetting, the work was pronounced a 'first-rate job'.[18]

It was while in Bramah's employ that Maudslay's mechanical skills were able to flourish. The major challenge facing Maudslay was the development of machines to facilitate the commercial production of Bramah's patent lock. Bramah had been able to construct his invention using hand skills but the labour costs were prohibitive. What was needed was the construction of specialised contrivances to replace the hand skills with mechanical processes. As Rolt has pointed out, the fact that the patent was taken out in 1784 and that commercial production did not begin until Maudslay joined him six years later gives some indication of the importance of Maudslay's role.[19] It was Maudslay who was able to translate Bramah's ideas into practical form and this was achieved as a result of his superior craft skills and mechanical ingenuity. The resulting tools included a sawing machine, a spring-winding machine and a quick grip vice, all of which are now in the Science Museum. Commenting on the construction of these and other lock-making machines, John Farey, writing in 1849, referred to 'a systematic perfection of workmanship which was at that time unknown in similar mechanical arts'.[20]

By the time Maudslay left Bramah's in 1797 his reputation as a master craftsman was secure, but it was soon to acquire further confirmation following a series of meetings with Marc Brunel. While in America escaping the extreme republicanism of revolutionary France, Brunel had devised his scheme to mass produce ship's pulley blocks. Shortly after landing in England in 1800, Brunel was introduced to Maudslay by a French emigré who had been impressed with the latter's work, especially his examples of screw-cutting. Brunel first employed Maudslay to construct models of his block-making machinery. These served to impress Sir Samuel Bentham, Inspector-General of Naval Works, who recommended to the Admiralty that Brunel's block-making machinery be installed at the Portsmouth Dockyard. In 1802 Maudslay was engaged to construct this machinery, a task that was to take seven years. In total there were forty-five machines, including sawing machines, boring machines, mortising machines, shaping engines and pin-turning lathes. Most were made entirely of metal and many remained in use for nearly 150 years. They were masterpieces of construction and one 1824 account stated that 'there is but another workman, perhaps, in the United Kingdom, who could have

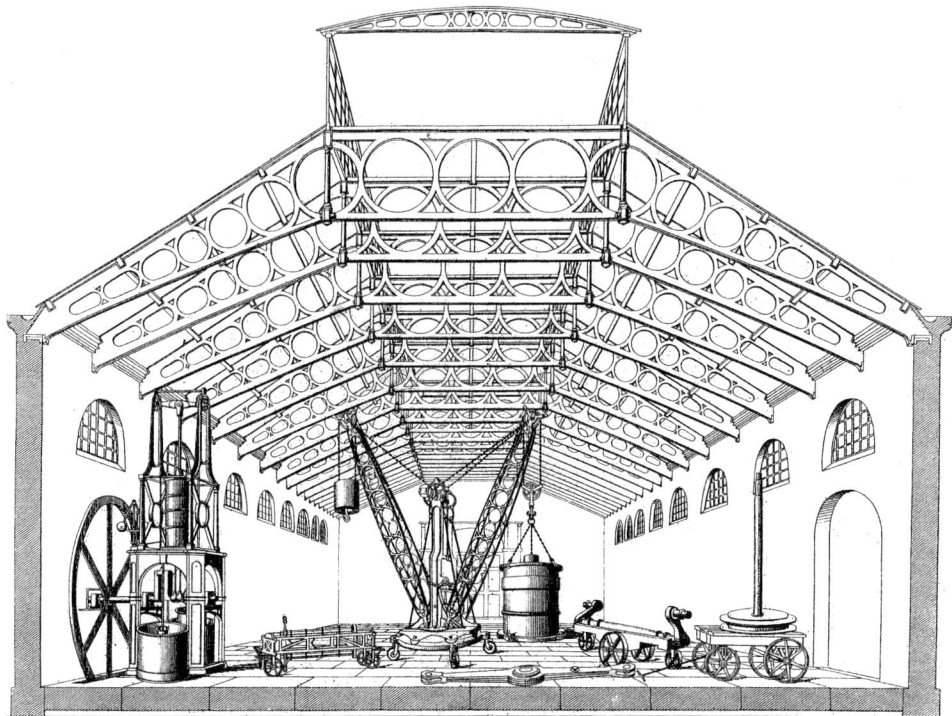

Fig.5 Henry Maudslay's erecting shop, 1830. *(Bulletin de la Société d'Encouragement pour l'Industrie Nationale, Paris, 1834)*

finished the engines in a manner so worthy of the invention'.[21] Joshua Field, in his Presidential Address to the Institution of Civil Engineers in 1848, said 'The design and execution of these machines reflect almost equal credit upon the inventor and the constructor; and if they are examined at the present time, they will be found not only unrivalled in workmanship; but embodying most of the important elements of the improved tools and machines of the present day'.[22] The blockmaking machinery extended the fame of Henry Maudslay and ensured his place as one of the most popular machine-makers in London.

Maudslay entered the patent lists just six times. Four of his patents were obtained before the move to Lambeth Marsh: in 1805 and 1808 he patented improvements in calico-printing machinery;[23] in 1806, together with Bryan Donkin, he patented a differential motion for raising weights;[24] and in 1807 he patented his 'table engine'.[25] The two remaining patents reflected Maudslay's interest in marine engineering: a method of purifying ship water by aeration to prevent the liquid from becoming 'tainted' or 'stinking' in the casks, a joint patent with Robert Dickinson dated 1812;[26] and a method of regulating the supply of water to boilers at sea and preventing the formation of brine in the boilers, a joint patent with Joshua Field dated 1824. This latter process was intended to improve the efficiency of boilers since

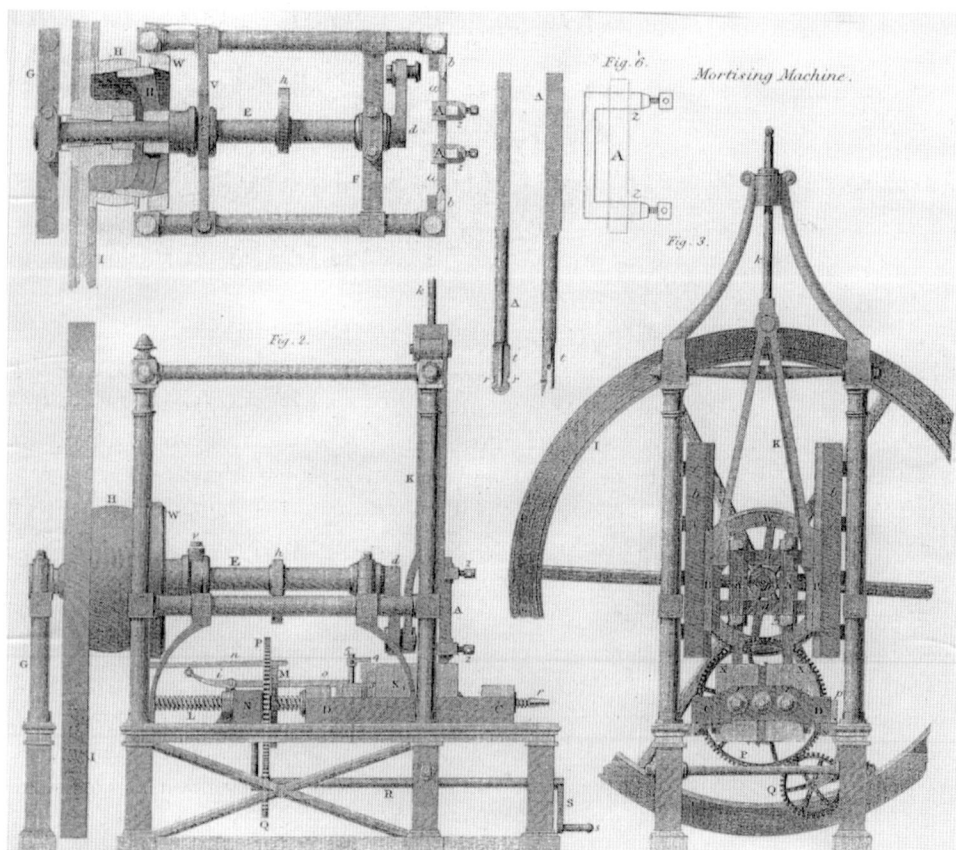

Fig.6 Mortising machine.*(Rees)*

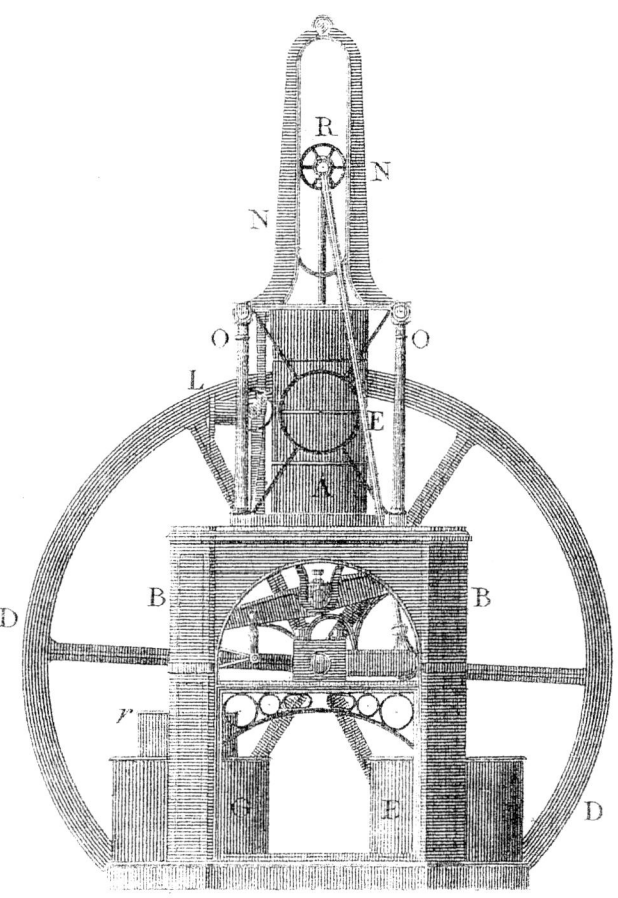

Fig.7 Table engine.*(Rees)*

salted water was 'unfit for raising steam'. It was also intended to increase the efficiency of steam ship travel in general by making unnecessary the stops to empty and refill the boiler every fifty or sixty hours (this was done to avoid the risk of serious boiler damage from salt deposits).[27] In addition to these patented inventions, Maudslay designed important improvements in coining and gun-boring machinery and invented a machine for punching holes in boiler plates which reduced the cost of preparing them for riveting from 7s to 9d a tank.[28]

The best known of Maudslay's patented inventions was his pyramidal or table engine, so called because the cylinder is positioned vertically on a cast iron table which also provides the support for the axis of the fly wheel. The beam and condensing cylinders are placed beneath the table top and, in what soon became a modification of the original specification, two long connecting rods attached to the guided crosshead at the top of the piston rod descend past the cylinder through slots in the table to the crankshaft.[29] The advantages of this arrangement are explained in Rees's *Cyclopaedia*: 'The specification states the invention to consist in reducing the number of the parts of the common steam engine, and so arranging and connecting them, as to render it more compact and portable, every part being fixed to, and supported by, a strong frame of cast-iron, perfectly detached from the building in which it stands: it is not, therefore, liable to be put out of order by the sinking of the foundations'.[30] Nasmyth emphasised the 'great simplicity' and 'get-at-ability of parts' and claimed that its 'compactness and self contained steadiness has caused it to be succeeded by a vast progeny all more or less strongly marked with the distinguishing features of the parent design and is still in as high favor as ever'.[31] The main purpose of the table engine was to drive factory machinery, and three such engines were installed at the Lambeth Works between 1815 and 1825. Competitive prices were achieved by manufacturing the engine in a number of standard stock sizes – a 6hp engine was delivered to the Woolwich Arsenal in 1809 for £645 and an 8hp engine was sent to the Conservatoire des Arts et Métiers, Paris, in 1814.[32]

Two inventions are attributed to Maudslay by James Nasmyth: the self-tightening collar for Bramah's hydraulic press, and the slide rest. In a letter to Samuel Smiles in January 1863, Nasmyth wrote:

> *Maudsley told me or led me to understand that it was he who invented the self tightening leather collar for the Hydraulic press without which it never would have been a serviceable machine, as the self tightening Leather collar is to the Hydraulic press so is the Steam Blast to the Locomotive. It is the one thing needfull that has made it effective in practice. If Maudsley was the inventor of the collar that one contrivance ought to Immortalize him – he used to tell me of it with great gusto and I have no reason to doubt the correctness of the naration* [sic].[33]

Joseph Bramah's hydraulic press was patented in 1795[34] and consisted of a large cylinder housing an accurately fitted plunger or ram. Small quantities of water were pumped into the cylinder beneath the ram so causing it to rise up to enforce compression. The problem lay in making a joint so that the cylinder was watertight, yet still allowing the ram to move freely. As shown in the first of the two rough sketches by Nasmyth,[35] Bramah's first idea was to use a stuffing box comprising a coil of hemp on leather washers placed in a recess X X held in place by a compressing collar forced down by strong screws. One problem with this arrangement was that the grip was so tight that the ram would not return downwards after the water pressure had been removed. Maudslay devised a way in which the water itself would give the necessary watertightness to the collar, by inserting an inverted U sectioned leather collar into the recess Z Z as shown in the second sketch. The result was that when the high pressure water entered the cylinder 'it forced its way into the leathern concavity and "flapped out" the bent edges of the collar' causing 'the leather to apply itself to the surface of the rising ram'.[36] When the pressure was removed the collar collapsed so allowing the ram to slide down.[37]

Maudslay's role in the development of the slide rest has often been misunderstood. Despite the ambiguous comments of Nasmyth and Smiles he was not the inventor of this mechanism. A form of slide rest can in fact be traced back to the fifteenth century.[38] More recently it was used by French ornamental turners during the eighteenth century and on Dutch imported gun-boring mills at the Woolwich Arsenal from the early 1780s.[39] It is highly probable that Maudslay saw these machines while working in the smith's department at the Arsenal, and in 1794 he devised his own slide rest while working for Bramah. Significantly Maudslay did not himself claim to have invented the slide rest, neither did Joshua Field make this claim for him in his brief summary of Maudslay's achievements delivered before the Institution of Civil Engineers in 1848. What Field does say on the subject is that Maudslay 'introduced the

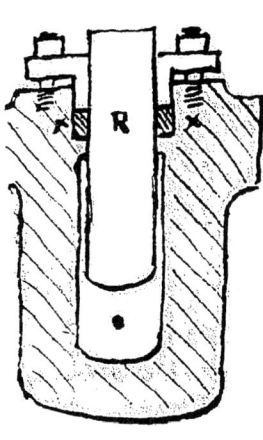
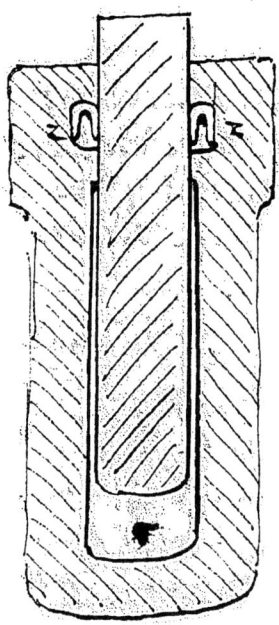

(a) Bramah's original idea
using a stuffing box.

(b) Maudslay's improvement
using a self-tightening collar.

Fig.8 Nasmyth's sketches to show the importance of Maudslay's self-tightening collar for the hydraulic press. *(Nasmyth Manuscript)*

general use of the sliding rest throughout his manufactory'.[40] It appears to be the case that Maudslay quickly appreciated the importance of the slide rest for achieving new levels of accuracy, and facilitating standardised production, and therefore insisted on its deployment throughout his works. This dedicated promotion of an existing mechanism which, in the late eighteenth and early nineteenth centuries, was all but unknown in British industrial workshops, was one of Maudslay's principal achievements. But while he refined, improved[41] and promoted the slide rest he did not invent it.

The popular legend to the contrary may have originated from 1806 with Olinthus Gregory who claimed that for 'turning faces of wheels, hollow work, &c. where great accuracy is wanted, Mr *Maudslay* has contrived a curious apparatus which he calls a *slide-tool*'.[42] The principal promoter of the view that Maudslay invented the slide rest, however, was James Nasmyth, though he never actually said that Maudslay was the inventor. Nevertheless, his *Remarks on the Introduction of the Slide Principle* published in 1841 could easily be misinterpreted. Nasmyth wrote: 'It would be blameable indeed … were I to suppress the name of that admirable individual to whom we are indebted for this powerful agent towards the attainment of mechanical perfection. I allude to the late Henry Maudslay … to him we are certainly indebted for the slide rest.'[43] Nasmyth later submitted to Smiles the view that Maudslay was 'the Parent of its introduction into practical service' though he also conceded that 'it would

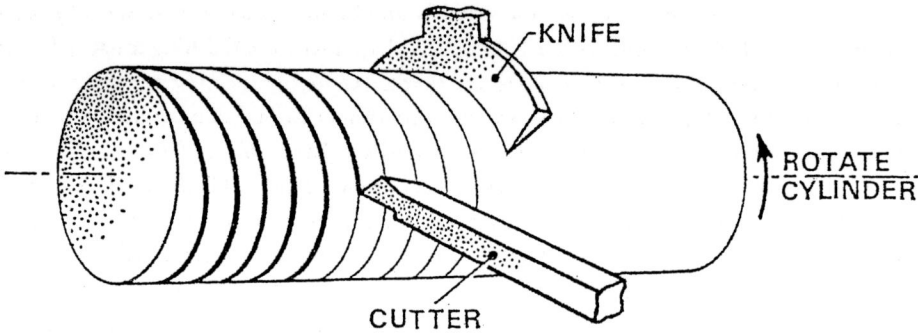

Fig.9 Screw-generating machine. *(Gilbert, 1971-1972)*

be giving him too much credit to say he was the inventor of the slide rest'.[44] Smiles, perhaps under the impression that Nasmyth was hair-splitting, refers to the slide rest as 'Maudslay's invention' though he also states that Maudslay 're-invented' the mechanism.[45] It is hardly surprising therefore that later writers became confused and made exaggerated claims for Maudslay.[46] Even Richard Prosser, the writer of Maudslay's entry in the *Dictionary of National Biography*, leaves the explanation of the slide rest's origins to Nasmyth, reproducing his 'Remarks' of 1841 verbatim.

James Nasmyth's *Autobiography*, published in 1883, made a claim for Maudslay that was omitted from Field's brief assessment and from the chapter in *Industrial Biography*: that he produced accurate standard planes. These, in the form of surface plates, were used as guides by Maudslay's workmen and were essential for the surfaces of steam engine valves, the production of lathe beds, and the structures of other machine tools and machinery such as the tables of printing presses. Maudslay's standard planes were made three at a time so that 'by the mutual rubbing of each on each the projecting surfaces were effaced'.[47] Any residual projections were removed by hard steel scrapers. When placed over each other these plane surfaces would float on a thin cushion of air and when they adhered to each other they could only be separated by sliding off each other. A recent biographer of Whitworth[48] has questioned whether Nasmyth was attributing to Maudslay what was, in fact, the achievement of Whitworth, as explained by the latter to the Glasgow meeting of the British Association in 1840.[49] There is certainly a striking similarity between Whitworth's description of his own methods to achieve true plane surfaces and those attributed to Maudslay. The evidence tends to suggest that while it is perfectly consistent with Maudslay's workshop practices that he should appreciate the importance of true plane surfaces, the standards of accuracy he actually achieved were imperfect and soon outdistanced by his former pupil. Indeed, since Whitworth began working on the problem while employed by Maudslay, it is quite likely that Maudslay set him the challenge of achieving a new level of accuracy. Why Nasmyth should seek to deprive Whitworth of the credit in this matter more than forty years after the Glasgow meeting may have been a slip of the memory or, alternatively a deliberate attempt to downgrade Whitworth, if stories of antipathy between the two are to be believed.[50]

Another area where Maudslay and Whitworth worked together was in the production of standard screws. Nasmyth wrote that Maudslay was 'the first to introduce system into screws which has been so admirably followed out to perfection by the subsequent labours of Joseph Clement and Joseph Whitworth who may be said to have given the finishing touches to our present perfect system of screws'.[51] Nasmyth also wrote that the production of perfect screws was one of Maudslay's 'highest ambitions and his principal technical achievement'. Nasmyth's claim is corroborated by Joshua Field, Samuel Seaward and James Walker.[52] Before Maudslay, screws were mostly cut by hand using a combination of file, chisel and hammer. This process was not only expensive, leading to the substitution of cotters, cotterils and forelocks, but it prohibited the introduction of standardisation. Every bolt and nut manufactured was unique and had to be marked as belonging to each other. Maudslay's achievement was to introduce standardisation within his own works, whereas Whitworth established an internationally recognised standard.

Perfect screws were necessary for accuracy of construction, accuracy of measurement and the promotion of interchangeability and mass production. Maudslay keenly appreciated the limitations of engineering manufacture without a regular sytem with regard to the ratio of the pitch or numbers of threads to the inch or the form of the threads themselves. According to Charles Holtzapffel writing in 1846, 'The late Mr Henry Maudslay, devoted an almost incredible amount of labour and expense, to the amelioration of screws and screwing apparatus; which, as regarded the works of the millwright and engineer, were up to that time in a very imperfect state'. In order to attain screws of exact values 'he employed numerous modifications of the chain or band of steel, the inclined knife, the inclined plane, and indeed each of the known methods...'[53] It appears that it was by the 'inclined knife' that Maudslay was most successful for his screw-generating machine, devised around 1800, embodied this principle and became 'the parent of a vast progeny of perfect screws, whose descendants ... are to be found in every workshop throughout the world...'[54] This machine, now in the Science Museum, consisted in the deployment of an adjustable steel knife which cut into the surface of a revolving bar of soft metal so forming a spiral groove. This groove was then deepened by a steel tool fixed in an adjustable slide.[55] Once a satisfactory screw had been made by this means it could be used as the master in a screw-cutting lathe to produce other screws in harder metal. Holtzapffel claimed that 'during the period from 1800 to 1810, Mr Maudslay effected nearly the entire change from the old, imperfect, and accidental practice of screw-making ... to the modern, exact and sytematic mode now generally followed by engineers; and he pursued the subject of the screw with more or less ardour, and at an enormous expense, until his death...'[56] One device made possible by the production of an accurate screw

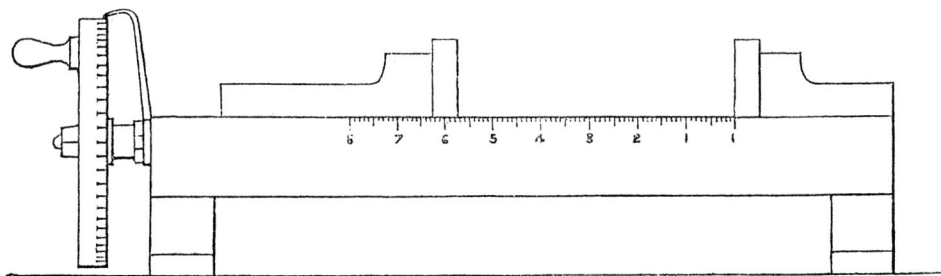

Fig.10 Bench micrometer. *(Nasmyth, 1883)*

Fig.11, above: Screw-cutting lathe, c.1797. *(Science Museum/ SSPL)*

Fig.12, left: Screw-cutting lathe, 1800. *(Science Museum/ SSPL)*

was Maudslay's bench micrometer or 'Lord Chancellor' – so named because it was used to settle all disputes about dimensions in the Lambeth workshops. The lead screw of one hundred threads per inch was used to move two steel blocks against the object to be measured and levels of accuracy to one ten-thousandth of an inch could be read from the scales.[57] Perhaps Maudslay's most impressive achievement in connection with the production of accurate screws was a screw 5ft long and 2in in diameter cut with fifty threads to the inch; the accompanying nut was one foot long and contained six hundred threads. This screw, which far outstripped the standards of accuracy necessary for the industrial workshop, was to be used at Greenwich for dividing the scales of astronomical instruments. It was sent for exhibition to the Society of Arts and won a government award of £1000.[58]

Armed with the slide rest, true plane surfaces and accurate screw threads, Maudslay was able to improve the construction of the lathe.[59] Prior to Maudslay the industrial lathe was a hybrid construction in terms of materials, with wooden framing encompassing iron parts such as the tool rest and centres. Such a method of construction failed to provide the rigidity necessary for the production of accurate work. All the Maudslay lathes were of metal construction and the noted elegance of design was achieved in part by his deliberate avoidance of sharp interior angles. A number of Maudslay lathes have survived, three of which are in the Science Museum. The first of these, dated around 1797, is a screw-cutting lathe fitted with a slide rest complete with a graduated dial to enable the depth of the cut to be determined. One unusual feature of the machine is the lead-screw mounting which is designed so that different lead-screws can be fitted. The 3ft bed consists of two triangular bars.[60] When the American engineer G.E. Sellars visited Lambeth in 1832 he described this machine as 'the father of all lathes, that by a combination of gear wheels and one guide screw any variety of pitch could be produced'.[61] A second hand-operated screw-cutting lathe bearing a brass plaque inscribed 'H. Maudslay 1800' is of instrument-maker size. With this lathe the lead-screw is permanently mounted within the bed, and the slide rest carries an adjustable stay to prevent any springing of the rod that might result from the thrust of the cutting tool. The third machine, a bench or treadle lathe, has a triangular bar bed and can be adapted for screw cutting. The upper slide of the slide rest is carried on a swivel and can be set for any variety of angles. It was manufactured for sale to amateur mechanics at a price of approximately £200. A much larger lathe with an industrial or manufacturing capacity is now kept at the Henry Ford Museum, Dearborn, Michigan. This machine which has a bed of 9in and a swing of 12in, is thought to date from around 1805. It represents a scaling up of the accuracy achieved in Maudslay's original lathe while maintaining the elegance of design.[62]

Maudslay's primacy in lathe design is well attested by contemporaries. Olinthus Gregory, writing in 1806, stated that 'the improved lathes manufactured by Mr. *Henry Maudslay*, of Margaret-street, Cavendish-square, are the most curious as well as the most useful of any we have seen'.[63] The entire section in Rees's *Cyclopaedia* on the lathe, written by John Farey, is devoted to a Maudslay treadle lathe decribed as 'the most perfect of its kind'. Farey states that Maudslay 'has a great number of different sizes, but on similar construction, in constant use at his manufactory for steam-engines and other machinery in the Westminster-road'.[64]

Some idea of Maudslay's approach to engineering design can be deduced from the various maxims he imparted to the young James Nasmyth, who faithfully recorded much of what his master said to him. One favourite piece of Maudslay advice was 'First, get a clear notion of

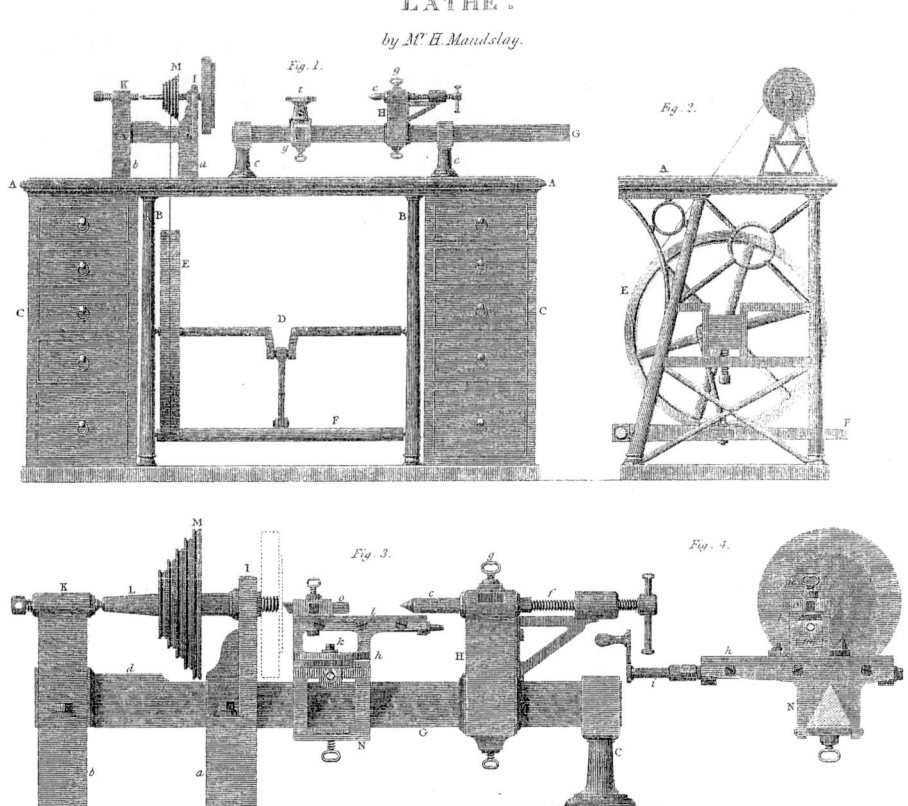

Fig. 13 Bench Lathe. (Rees)

what you desire to accomplish, and then in all probability you will succeed in doing it'. Another was 'Keep a sharp look-out upon your materials; get rid of every pound of material you can do without; put to yourself the question, "What business has it to be there?" avoid complexities, and make everything as simple as possible'.[65] Further homilies included 'There is a right way and a wrong way of doing everything'[66], and 'When you want to go from London to Greenwich, don't go round by Inverness'. Maudslay considered no man a thorough mechanic 'unless he could cut a plank with a gimlet, and bore a hole with a saw!'[67] An observation of more practical value was 'We must not become too complicated with our machinery. Remember the get-at-ability of parts. If we go on as some mechanics are doing, we shall soon be boiling our eggs with a chronometer!'[68] Clearly Maudslay favoured meticulous planning and preparation to achieve simplicity, economy and practicality of design. Such are the unmistakeable hallmarks of Maudslay's inventions and improvements.

James Nasmyth has left a portrait of Maudslay's personal appearance and character:

> *If ever the Beau Ideal of John Bull was seen it was in Henry Maudsley his height 6ft 2in and massive portly frame lighted up with his handsome jolly countenance and good natured but most penetrating eye [.] his hearty laugh and thorough good natured yet attractively dignified aspect*

would cause anyone to be charmed by him. There was a genial frankness about him that soon set those at their ease who approached him for the first time. no one could be more faithfull and constant in his attachments [.] he was surrounded by old friends who delighted to come and sit by him while he was at work in his own beautifull little private workshop in which he spent most of his time carrying out some new notion in mechanism and working at such with all the zeal as if he was beginning life [.] but while he was frank and cordial and kindly to the highest degree to those he had reason to love and respect he could turn what he termed his dark side towards those who had acted unworthily in such a way and with a few sarcastic words as was almost payment in full for all those misdoings [.][69]

Henry Maudslay died on the morning of 15 February 1831. While in the midst of an enthusiasm for constructing an observatory at Norwood, he was summoned to visit an ailing old friend at Boulogne. He remained there for a week until assured of his friend's convalescence. On the return trip across the Channel he caught a cold so severe that he took to his bed upon reaching London. He never left it alive.[70] Maudslay made codicils to his will on 10 January and 3 February 1831 so he may have had some intimation that his end was close. Among other bequests he left £5 to every workman or servant who had been in his continuous service for twenty years. Maudslay was buried in the churchyard of St Mary's, Woolwich, overlooking the River Thames. The cast iron tomb of his own design was cleared away by

Fig.14 Industrial Lathe c.1805. *(From the Collections of Henry Ford Museum & Greenfield Village)*

local authority workers in 1966 to make way for a landscaped garden.[71] Fortunately photographs and transcriptions have ensured that the text of Maudslay's epitaph has survived:

> TO THE MEMORY OF
> HENRY MAUDSLAY
> BORN IN THIS PARISH 1771
> DIED AT LAMBETH FEB 15TH 1831
> A ZEALOUS PROMOTER
> OF THE ARTS & SCIENCES
> EMINENTLY
> DISTINGUISHED AS AN
> ENGINEER
> FOR MATHEMATICAL
> ACCURACY AND
> BEAUTY OF CONSTRUCTION
> AS A MAN FOR
> INDUSTRY & PERSEVERANCE
> AND AS A FRIEND FOR A KIND &
> BENEVOLENT HEART

Few students of Maudslay would take issue with these words of appreciation, probably composed by Joshua Field. But Maudslay's historical importance is greater than this inscription suggests, as many writers on the history of technology have acknowledged. Rolt argued that 'mechanical engineering technique was completely revolutionised by his example' and that he began a 'transformation' from a hand craft technology to one based on automated machines.[72] Petree spoke of Maudslay as 'the engineers' engineer' whose distinguishing legacy was 'to ensure and maintain not merely accuracy, but repeatable accuracy in workshop production'.[73] Evans described Maudslay's importance in terms of 'the bridge between the great craftsmen of the eighteenth century ... and the new machines where rigid frames, accurate screws, plane surfaces guided the tool'.[74] The idea of a man whose life and work spanned two worlds or cultures is critical to any proper understanding of Maudslay's significance. As with his 'Ideal Hero' Napoleon,[75] he became the acknowledged master of both the old and the new, a highly skilled craftsman who appreciated the limitations of those very skills and worked to embody them in mechanical mechanisms. Without wishing to detract from Maudslay's greatness, he was nevertheless the right man, in the right place, at the right time. Thirty years earlier, and he would have been remembered as a magnificent craftsman but without any far reaching influence over the development of the engineering profession. As it was, he was able to encompass both the achievements of the craft-based, eighteenth-century technology, and promote the machine-based technology of the nineteenth century. He was able to do this partly because of the length of his working career. Although he died comparatively young, at the age of fifty-nine, his working career spanned forty-eight years.

Maudslay was not the only great mechanical engineer of his time, for both Matthew Murray of Leeds and James Fox of Derby were successful inventors and manufacturers. But Maudslay was almost certainly the pre-eminent engineer of the age, as Gilbert and others have recognised.[76] While there were occasional criticisms of the company products,[77] it is difficult to argue with the plethora of engineering talent that drew its inspiration from Maudslay and continued to promote and practise the merits of his improved engineering methods and standards after his death. It is surely here that Maudslay's true importance lies, for he lived at a time when a number of key industries, notably textiles, transport and iron, underwent rapid technological change and expansion. Demand for steam engines, boilers, spinning and weaving machines, furnaces and forging apparatus, cranes, bridges, and all the inventions and improvements associated with the classic phase of the industrial revolution and the emerging machine age, placed a heavy responsibility on the suppliers of such machines and equipment. What was required was quantities of cheap, accurate and reliable engineering products. Maudslay's example provided the key to meeting this challenge, for it was through the deployment of his improved workshop practices that a bottleneck in the production of capital goods was avoided. This in turn enabled the manufacturing sector of the British economy to flourish for much of the nineteenth century, ensuring a prolonged period of industrial supremacy. Henry Maudslay, whose role in this process has often been overlooked, was truly a creator of the industrial age.

Endnotes

[1] James Nasmyth is the major source for Maudslay's life and work and he provided Smiles with the material for the 'Henry Maudslay' chapter in *Industrial Biography,* 1863. Further details are provided in Nasmyth's *Autobiography* (1883).The only full scale assessment of Maudslay is published in Russian: Zagorskii and Zagorskaia 1981. There are a number of good shorter assessments: see Roe 1916, Petree 1934, 1934-35, 1949, 1964, Rolt 1965, Gilbert 1971, 1971-72 and Evans 1994-95. There is a Maudslay archive in the Science Museum Library.

[2] The other five engineers were Rennie, Telford, Watt, Bramah and Brunel: Fairbairn 1859.

[3] *Behandlungen Des Vereins Zur Beförderung Des Gewerbfleisses In Preussen,* 12, 1833, p.248. The author of this account, a Mr Wedding, also commented that Maudslay's workshop 'is one of the most extensive and best equipped in London'.

[4] Biographical details of Maudslay's life and that of his father are based on the account in Smiles 1863, together with an undated manuscript prepared for Smiles by James Nasmyth entitled 'Some particulars of the life of the Late Henry Maudsley furnished by him viva voce to James Nasmyth in 1830': BL Add.Mss. 71075, ff.139-150.

[5] The change in spelling from Maudsley to Maudslay first occurred on Henry Maudsley's marriage certificate dated 26 July 1763.

[6] Information on Maudslay's origins is derived from a geneological table, dated 1880, compiled by Henry Maudslay (1822-1899) and Henri Marillier: SMAC, MAUD 15/2.

[7] J. Simister to J. Woodward, 31 October 1799: Boulton & Watt Collection, Parcel C/22.

[8] Rate assessment books list Maudslay at 76 Margaret Street in 1805 and at No.78 in 1810: St Marylebone Rate Book, City of Westminster Archives.

[9] *PICE,* 23 (1863-64), p.489.

[10] Statement of capital profits and withdrawals relating to H. Maudslay and J. Mendham, 1813-1820: SMAC, MAUD 15/2.

[11] Report on a Journey to England by Factory Commissioner J.G. May in 1814, quoted in Henderson

1968, p.158. May was a Prussian civil servant sent to England as part of his official duties to report on new machinery and industrial processes.

[12] *PICE*, 69 (1881-82), p.411.

[13] Gilbert 1971, p.26.

[14] An advertising circular watermarked 1812 refers to 'Henry Maudslay & Co.': SMAC, MAUD 7/1. Critchett and Woods' *London Directory* for 1813 refers to 'H. Maudslay & Co.'

[15] *London Gazette,* 17661,19 December 1820, p.2383.

[16] Will of Henry Maudslay: SMAC, MAUD 3/8.

[17] Allen 1826, p.319. In May 1826 Maudslay reportedly had 500 men in his employ: *The Times,* 25 May 1826, p.3.

[18] Nasmyth Manuscript ff.140-142. See note 4.

[19] Rolt 1965, p.85. A contrary view has been argued by Dickinson, supported by Bramah's biographer: Bramah is unlikely to have gone to the trouble of publishing his *Dissertation on the Condition of Locks,* 1788, if he had been unable to manufacture the locks at that time: Dickinson 1941-42 and McNeil 1968, p.48.

[20] *PICE*, 9 (1849-50), pp.331-32.

[21] *Encyclopaedia Britannica,* Supplement to fourth edition, 2 (1824) p.338. For a full account of the blockmaking machinery, see Gilbert 1965.

[22] *PICE,* 7 (1848), p.37.

[23] Patents 2872, 1805 and 3117, 1808.

[24] Patent 2948, 1806. This mechanism was to replace the crane and could also be applied to driving screws, capstans and lathes.

[25] Patent 3050, 1807.

[26] Patent 3538, 1812.

[27] Patent 5021, 1824.

[28] Smiles 1863, p.224. For a description of this machine by Maudslay's grandson, Henry (1822-1899), see *PICE*,17 (1857-58), pp.193-94.

[29] For a detailed description of the model of Maudslay's table engine housed by the Science Museum see Gilbert 1971, pp.21-22. In the patent specification two long side rods were attached to the crosshead at the top of the piston rod which connected with a guided grovel wheel at the base of the table between the two cast iron water cisterns. Two connecting rods linked this wheel with the crank bearings.

[30] Rees, 5 (1819-20), p.131.

[31] Nasmyth Manuscript f.148. See note 4.

[32] Petree 1934-35, p.26 and Gilbert 1971, pp.21 & 25.

[33] J. Nasmyth to S. Smiles, 13 January 1863: BL Add. Mss. 71075, ff.15-16.

[34] Patent No.2045, 1795. There is no mention of the self tightening collar in the patent specification.

[35] Nasmyth Manuscript f.143. See note 4.

[36] Smiles 1863, p.190.

[37] Maudslay's claim to this invention is contested by the author of an article on the history of the hydraulic press who attributes it to Benjamin Hick of Bolton: *Mechanics Magazine*, XI, New Series, 3 June 1864, pp.360-61. Bramah's biographer questions whether the self-tightening collar was so revolutionary and also quotes a letter from Bramah's grandson that casts doubt on the claims of Nasmyth and Smiles: McNeil 1968, pp.111-117.

[38] Woodbury 1961, p.47.

[39] J. Anderson to General Lefroy, 18 March 1867, quoted in an appendix to the 1878 edition of *Industrial Biography*.

[40] *PICE*, 7 (1848), p.37.

[41] Maudslay's first slide rest (excepting the device made jointly with Bramah in about 1795 where the slide rest and tailstock were combined) was clamped to the lathe bed as in Fig.13; the tool could be moved right-to-left and in-and-out and vice versa by a short screw turned by a winch handle. If it was necessary to turn a bar the entire length of the lathe then the slide rest had to be unclamped, moved along the main lathe bed and then reclamped for a further session of work. Such a basic construction was adequate for rough work but quite inadequate for precision engineering. Hence Maudslay devised a compound slide rest moved by a lead screw geared to the lathe mandrel which could run the full length of the lathe bed as in Figs 11 and 12. Using this self-acting device a bar could be turned the entire length of the lathe with high levels of precision.

[42] Gregory 1806, p.472.

[43] Buchanan 1841, p.401. These claims were repeated in Baker 1858-59, p.232.

[44] Nasmyth Manuscript f.144. See note 4. It is interesting that in the extensive sections on Maudslay in Nasmyth's Autobiography there is no mention of the slide rest or of Maudslay's connections with the mechanism.

[45] Smiles 1863, pp.207 & 211.

[46] See, for example, Pole (ed.) 1877, p.41; 'Messrs. Maudslay, Sons and Field's Works', *The Engineer*, 52, 7 October 1881, p.252; Vincent 1888-90, p.215; 'Thames Pioneer Shipbuilders and Marine Engineers', *The Engineer*, 87, 27 January 1899, p.81; Sharp 2000, p.75. The text alongside the Maudslay stained glass portrait in Woolwich Town Hall states that he 'invented the Slide Rest for the Lathe.'

[47] Nasmyth 1883, pp.148-49.

[48] Atkinson 1996, pp.86-91.

[49] Whitworth 1840.

[50] For the alleged personal feud between Nasmyth and Whitworth see Atkinson 1996, pp.41, 43 & 88. Interestingly, in the contest between Nasmyth and Whitworth, most historians pre Atkinson, obviously unaware of the possibility of a mutual antipathy between the two men, side with Nasmyth who promotes the position of Maudslay : Roe 1916, pp.44-45; Petree 1964, pp.103-104; Musson 1975, pp.117-122.

[51] Nasmyth Manuscript, ff.147-148. See note 4.

[52] *PICE*, 1 (1841), p.160.

[53] Holtzapffel 1846, 2, p.641.

[54] Nasmyth 1883, p.140.

[55] For a full description of this device see Nasmyth 1883, pp.139-140, and Gilbert 1971, pp.15-17.

[56] Holtzapffel 1846, 2, p.647.

[57] For detailed descriptions of the micrometer see Nasmyth 1883, pp.149-150, and Gilbert 1971, pp.13-14. Maudslay was not the inventor of the micrometer, for which credit is due to William Gascoigne and James Watt, but he did provide new standards of accuracy. For an enlarged photograph of the thread of the 'Lord Chancellor' now in the Science Museum see Evans 1994, p.158. According to Petree (1949, pp.13-14), this machine was 'put through tests by the National Physical Laboratory in 1918 to determine its degree of accuracy, and, while it was found that, owing to a certain springi-

ness and want of rigidity, there were minute errors on small lengths, yet the degree of accuracy was quite extraordinary, considering the date of its construction'.

[58] Smiles 1863, pp.226-27.

[59] For Maudslay's contribution to the lathe and screw cutting see 'Some Early Machine Tools', Parts I, II & III, *The Engineer*, 97 (6 & 20 May and 3 June 1904), pp.453-54, 505-506 and 553-55.

[60] There are full descriptions of the Science Museum Maudslay lathes in Rolt 1965 and Gilbert 1966, 1971 and 1975.

[61] Ferguson 1965, p.111.

[62] I am grateful to Marc Greuther, Curator of Industry at the Henry Ford Museum, for supplying me with information about the sole surviving Maudslay industrial lathe.

[63] Gregory 1806, II, p.471. Gregory noted a number of large lathes in Maudslay's manufactory, worked by hand, together with some additional apparatus for cutting the teeth of wheels.

[64] Rees 1819-20, 3, p.249.

[65] Nasmyth 1883, pp.130-1.

[66] Nasmyth 1883, p.147.

[67] Nasmyth 1912, p.143.

[68] Nasmyth 1912, p.146. Nasmyth added further Maudslay maxims to the later editions of his *Autobiography*. The first edition attributes the 'get-at-ability' maxim to Joshua Field.

[69] Nasmyth Manuscript, f.149. See note 4. There are very few other sources for Maudslay's character though a newspaper reporter at the time of a serious accident commented that Maudslay 'is a very kind and humane man, and extremely attentive to the comforts of those in his employment': *The Times*, 25 May 1826, p.3.

[70] Nasmyth 1883, p.175.

[71] The tomb plate containing the epitaph to Henry Maudslay has apparently been lost or destroyed. The three remaining plates were transferred to Greenwich Borough Museum in August 1993.

[72] Rolt 1965, pp.90-91.

[73] Petree 1964, p.100.

[74] Evans 1994, p.172.

[75] Nasmyth Munuscript, f.149. Nasmyth wrote, '"Napoleon the Great" was his Ideal Hero and he spared neither trouble nor cost to get together every work of art that bore the impress or illustrated the greatness of that great man'. Maudslay was principally impressed by Napoleon's 'instigation of magnificent works.'

[76] Gilbert 1971, p.30, and 1971-72, p.50.

[77] Lt William Nairn Forbes, an employee of The East India Co., complained 'heavily' of the Maudslay engines sent out for the Calcutta Mint 'all the parts being too slight for the work': extract of a letter from P. Ewart to J. Watt Jr, 25 January 1829, Boulton & Watt Collection Box 19/5. Some eighteen months after Maudslay's death his works were visited by an American, George Escol Sellers, who commented on the inadequacy of the lathes with regard to 'weight, strength and firmness': Ferguson 1965, p.112.

3
The London Engineering Industry at the time of Henry Maudslay

A.P. Woolrich

Henry Maudslay's career reflects the development of the more progressive sector of the London engineering industry between the 1770s and 1830s.[1] Significantly he began his working life at the Woolwich Arsenal, the only major engineering enterprise where machine tools such as boring engines, drills and lathes were employed at the beginning of this period. Elsewhere, all engineering practice was rooted in a craft-based technology. By the turn of the century, Maudslay was one of a select group of highly skilled workshop engineers, including Joseph Bramah, John Rennie and John Penn, who collectively established London's reputation for engineering excellence. This was achieved mainly through the deployment of improved machine tools, though craft techniques such as chiselling and filing were still necessary to supplement the work of the new machinery. Some early mechanical engineers including Bryan Donkin and Arthur Woolf acquired factory premises during the early 1800s, but most did not acquire sufficient capital resources to expand in this way until the second decade of the nineteenth century. Maudslay began operating from his Lambeth works in 1810, and by the time of his death in 1831 was a senior partner in one of the largest and most successful engineering enterprises in the country, manufacturing a wide range of products from marine steam engines to gentlemen's benchtop lathes. By 1830 the metropolitan engineering entrepreneurs were using a more complete machine technology than that employed a generation earlier, with slide lathes, planing machines and gear-cutting engines representing the most important advances. But tools to perform specialist work were yet to be devised and craft skills were still needed for fitting and work of a specialist character such as the creation of small accurate flat surfaces and the cutting of slots.[2]

London's position as the principal centre of the engineering industry during Maudslay's lifetime resulted partly from the engineering needs of the largest concentration of population in the world. This mass market, which numbered about one million in 1801 and had nearly doubled by 1831, had to be fed, watered, clothed, and provided with sanitation, security, coinage, transport, lighting and recreation. All these needs had implications for machine builders and associated metal workers who were required to equip flour mills and breweries, to construct water closets, to manufacture piping, locks, fire engines, printing and engraving machines and all the more esoteric demands of the court and aristocracy such as fountains, carriages and wrought iron gates and railings. London had other important engineering-related functions. It was the nation's leading port[3] and centre for both coastal shipping and inland distribution; the road network connected London with all parts of the country while the Grand Junction Canal provided a central artery to the industrial midlands. This had the effect of extending the market for London engineering products and provided a number of specific demands. The on- and off-loading and storage of cargoes required cranes, pulleys,

Key

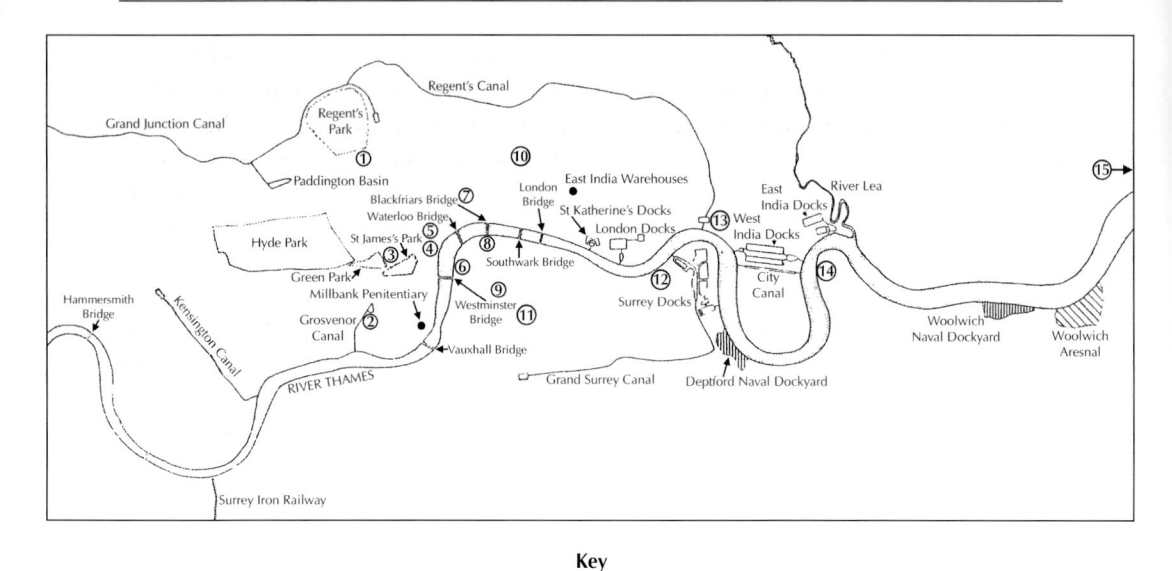

	Engineering firm	Location		Engineering firm	Location
1	Braithwaite	Marylebone	9	Maudslay	Lambeth
2	Bramah	Pimlico	10	Taylor & Martineau	Finsbury
3	Bramah	Piccadilly	11	Clement	Southwark
4	Maudslay	Oxford Circus	12	Donkin	Bermondsey
5	Holtzapffel	Covent Garden	13	Seaward	Millwall
6	Napier	Lambeth	14	Penn	Greenwich
7	Galloway	Holburn	15	Hall	Dartford
8	Rennie	Blackfriars			

Fig.15 Map to show location of principal engineering firms at the time of Henry Maudslay.

winches and hoists. The refining and processing plants for imported sugar, tobacco and hides located along the banks of the Thames needed vats, presses and pumping engines. The advent of steam navigation made London a major centre for marine engineering during the first half of the nineteenth century.

London was in many ways ill-equipped to become an industrial city since she had virtually no water power, no local sources of coal or iron and no cheap land or labour. The capital had a major advantage, however, in the skills of its workforce. The Clerkenwell area, just to the north of the city, housed an army of craft workers – clock and watch makers; scientific, surgical and optical instrument makers; and trades making a host of small metal parts such as locks, keys, candle snuffers and harness fittings. Among the practitioners of these occupations were such famous names as Dolland, Ramsden and Harrison. It was from these and other craft trades such as the millwright, smith and carpenter that the first generation of mechanical engineers emerged. John Rennie and John Penn first trained as millwrights while Bramah, Woolf and Maudslay began their working lives in carpenters' shops. Indeed as late as 1825 the engineer John Martineau claimed that he would respond to a sudden upsurge in demand by recruiting from the depressed handicraft trades such as watchmakers and mathematical instrument makers. Men from these trades could 'with a very little practice' be brought to perform 'a great deal of the work' in an engineering factory.[4] The transferable skills of craft workers in wood and metal meant that as soon as demand developed for metal machines there was a ready pool of labour available.[5] In many ways the technology of the watch and clockmakers was similar to that of the early mechanical engineers. The former used wheel-cutting engines, lathes and drills, and would have been skilled in the art of fine filing. The principal difference was one of scale. Up-sizing to a factory operation would occur only when there was some guarantee that the considerable capital investment necessary would be amply rewarded.

At the time of Maudslay's birth in 1771 it was hardly possible to speak of an engineering industry. There were skilled mechanics and civil engineers practising in London – such as Thomas Yeoman who had fitted ventilating apparatus to both Houses of Parliament and the Drury Lane Theatre[6] – but they were few in number, possessed meagre capital resources and operated from small workshops where the power source would have been a hand wheel or treadle. Engineering activities were sometimes an offshoot of a seemingly unrelated craft; for instance John Pickering, inventor of the drop stamp in 1766, worked as a London toymaker.[7] The Society of Civil Engineers established in London in 1771 contained only a handful of mechanical engineers among its membership, and the most noted of them, James Watt, was not London-based, though he supplied steam engines to the capital. There were no privately-owned engineering factories in London during the last quarter of the eighteenth century. The only engineering enterprises of note were associated with armaments production and warfare – the Royal Arsenal at Woolwich and the naval dockyards at Woolwich, Dartford, Sheerness and Chatham.[8] The most advanced machine tools of industrial capacity to be found in London in 1771 were the heavy boring mills designed by the Dutch immigrant Jan Verbruggen, newly appointed Master Founder at the Royal Arsenal.[9] Verbruggen's mortar-boring mill was a horizontal engine in which the boring heads remained stationary while the gun was made to revolve by a horse gin. A contemporary drawing of this machine shows that the Arsenal engineers used set-squares, callipers and possibly profile templates to promote accuracy of production. This was also advanced by the massive timberwork of the bed which supported the boring bar so ensuring a high level of rigidity. The necessary precision was not achieved, however, until the introduction of all-metal machines in the early nineteenth century.[10] Nearly all machinery in the last quarter of the eighteenth century was made of wood with iron, brass and leather fittings. The explanation for this is that, while there were hand tools for working in both metal and wood, there were virtually no machine tools apart from a 'few ill-constructed lathes, with some drills and boring machines of a rude sort'.[11] To have built a machine with industrial capacity out of cast iron would have involved an inordinate amount of hand labour to chip, file and scrape the rough casting. It may also have been difficult to find an iron founder willing and able to undertake such one-off work at an affordable rate. So although the early construction of an all-metal lathe would have paved the way for the relatively inexpensive production of other industrial-scale machine tools, the huge costs involved and the uncertainty of demand seem to have acted as an effective block to progress until the very end of the eighteenth century. It was much easier and cheaper to work in wood. Hence even the lock-making machines constructed for Bramah by Maudslay in the 1790s were rooted in the wooden framed tradition.

By the turn of the century, London engineering had begun to move towards a machine shop technology, reflecting an increasing dependence on steam engines and other machinery demanded by local industry.[12] Most of the lathes would have been fitted with tool rests, enabling the turner to hold his tools against the workpiece, an operation requiring considerable skill and dexterity. The slide rest had just begun to appear in the most advanced machine shops such as that of Maudslay.[13] The more enterprising and successful engineering masters were operating from larger workshops and the senior representative of this group was Joseph Bramah (1749-1814).[14] Bramah was the son of a Yorkshire farmer and served his apprenticeship as a joiner and cabinet maker. He arrived in London in 1773 and probably established his manufacturing business in Denmark Street, St Giles, between 1778 and 1783. In 1784 he moved to 124 Piccadilly, West End, where he remained until acquiring factory premises in Pimlico in 1806.[15] By 1814 the Pimlico works employed about 100 men and comprised a machine shop, a foundry, a pattern shop and a model shop.[16] The workshops were 180ft in length and some 40ft wide. When Bramah died the business was continued by his two sons, Timothy and Francis. Bramah's engineering inventions were closely related to the social needs of the London population. He devised improvements to the water closet (he had manufactured 6,000 by 1797) and the household lock, and invented a beer pump, the fountain pen and the propelling pencil. Undoubtedly Bramah's most important engineering contribution was the hydraulic press patented in 1795, which had early applications in the woollen industry, for hay packing and for oil extraction. Bramah extended the hydraulic principle to cranes, lock gates and for driving machinery, more than fifty years in advance of Sir William Armstrong. Besides the young Maudslay, Bramah employed Woolf and Clement. To achieve his engineering success, Bramah made use of specialist machines in connection with his lock-making; was one of the first engineers to deploy a basic form of slide rest in an industrial context; and perhaps provided the inspiration for one of the most important tools of the machine shop with his wood planing machine supplied to the Woolwich arsenal in 1805.[17]

Another leading London engineer of the early nineteenth century was John Rennie (1761-1821),[18] whose reputation rests principally on feats of civil engineering concerning the construction of roads, bridges harbours and lighthouses. In fact, he was also a mechanical engineer of distinction and established an engineering factory in Blackfriars, completed by 1810. Rennie's reputation in this field was established when Boulton & Watt contracted his services for the design of millwork and machinery to be installed at the Albion Mills in 1784. Between this date and the building of the Blackfriars factory, Rennie operated from workshops on Jamaica Wharf, Upper Ground Street, off Stamford Street. When the flour mills burnt down in 1791 Rennie was able to buy up part of the site, and after buying adjoining property had the space to begin constructing his factory premises. Following Rennie's death the Blackfriars factory was taken over by his eldest son, George (1791-1866), while the civil engineering practice was continued by the second son, John (1794-1874). Rennie is credited with the invention of the centrifugal governor adopted by Watt, and it is also claimed that he originated the gantry crane. Machinery manufactured at his Blackfriars factory included steam dredging machines, minting machinery, threshing machines, rope-making machines, printing machines, and all types of mill. Rennie was the first engineer to use steam power for pile driving and was one of the earliest engineers to use ball bearings. Only after Rennie's death did his firm begin to make marine and other steam engines[19] and, in later years, locomotives. Rennie's expertise was clearly recognized by Boulton & Watt who ordered machinery from him for the Soho Foundry during the 1780s and 1790s, including

Fig.16, right: Joseph Bramah (1749-1814). *(Institution of Mechanical Engineers)*

Fig.17, below: Bramah's hydraulic press. *(Rees)*

Fig.18 John Rennie (1761-1821).
(Institution of Civil Engineers)

Fig.19 John Hall (1764-1836).
(Hesketh, 1935)

the rolling mill and polishing machinery, a vertical boring bar and drilling and boring machines. Rennie was therefore one of the first engineers to produce machine tools for sale.

Other mechanical engineers operating at this time included John Jacob Holtzapffel, John Hall (1764-1836), Bryan Donkin (1768-1855) and John Penn (1770-1843), all of whom founded successful engineering firms. Holtzapffel[20] was a German who established a tool making business in London around 1787. This concern was developed by his son, Charles (1806-1847), the author of a famous three-volume work on *Turning and Mechanical Manipulation* (1843). Holtzapffel's business records show that the firm produced an annual turnover of more than £10,000 by 1811 and sales in that year included some fifty lathes, ranging in price from £12 to £300, and two 'rose engines' which sold for over £400 each. Holtzapffel's appear to have concentrated on high quality and high priced ornamental turning lathes for the wealthy amateur, though they also manufactured a wide range of specialist hand-tools for various trades. John Hall[21] was apprentice to his own father, William, a millwright from Laverstoke, Hampshire. At the age of twenty-one he set up business as a smith in Lowfield Street, Dartford, but within five years needed larger premises which were found in Waterside on land which had once formed part of Dartford Priory. John Hall acquired a reputation for his work on paper-making machinery[22] but his versatility lead him into areas as diverse as steam engine manufacture, rolling mills, machinery for plate-glass works and hydraulic presses. One of his apprentices, and later brother-in-law and business partner, was Bryan Donkin.[23] Donkin moved to Dartford from Northumberland during the 1780s on the advice of John Smeaton, becoming indentured to Hall in March 1792 for a term of three years. He developed Hall's interest in paper-making machinery and in 1798 established his own business as a mould-maker, financed by Hall to the sum of £350. He then began work on the first practical paper-making machine on behalf of the Fourdrinier brothers who established a works at Bermondsey. They were bankrupted in November 1810 and in the following year Donkin obtained possession of the factory. Donkin also turned his attention to printing machines, the design and construction of water wheels, and food-canning – he acquired Peter Durand's patent for the preservation of animal and vegetable foods and formed a company to exploit it together with Hall and John Gamble. The youngest figure in this group, John Penn,[24] was born near Taunton and apprenticed to a millwright at Bridgwater, Somerset. He moved to London in about 1793 and in 1800 started a small millwright's and machinist's shop at Greenwich. Here Penn was involved with the manufacture of windmills, waterwheels, treadmills and the accompanying machinery. His firm did not become involved in the production of marine engines, a line of business that secured his reputation, until the mid-1820s.

There were many other mechanical engineers of note working in London during the early years of the nineteenth century such as Arthur Woolf (1766-1837),[25] John Braithwaite (d.1818),[26] and William Dixon,[27] but in many cases there is virtually no surviving evidence about their activities. The firm of Coulson, Jukes and Pelly is now quite unknown yet J.C. Dyer listed it as one of the most prominent engineering firms in London during the 1800s.[28]

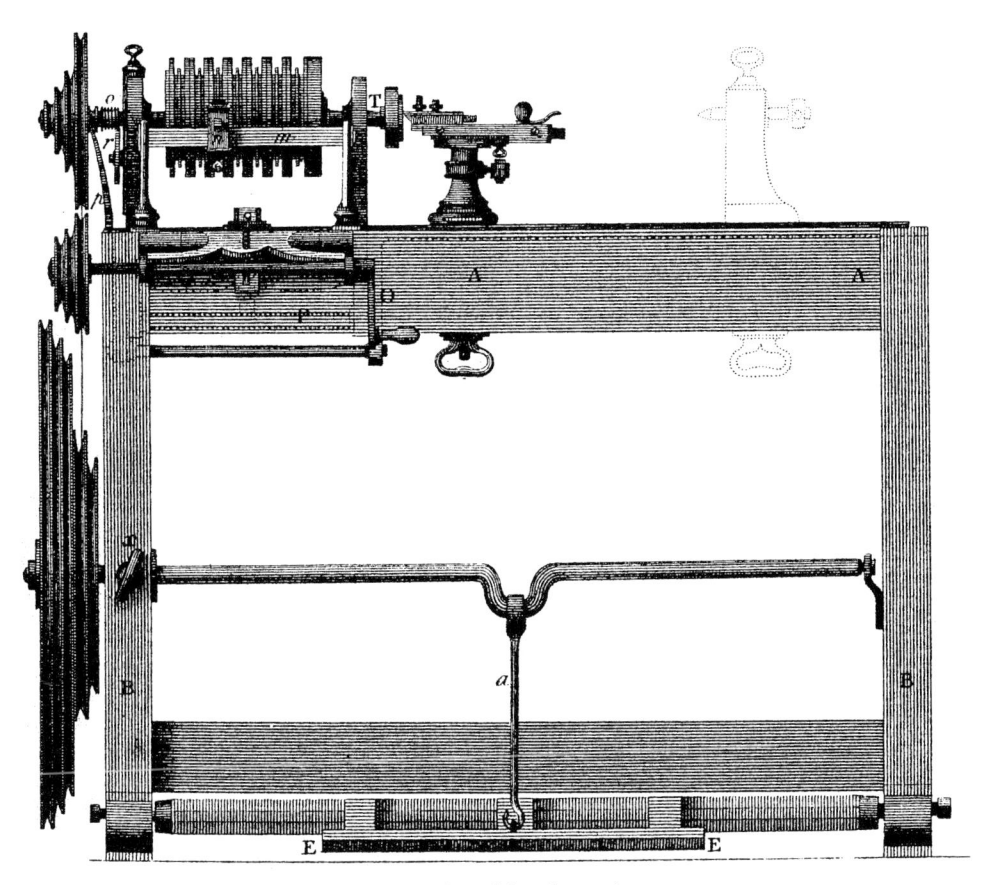

Fig.20 Rose engine or figure lathe made by Holtzapffel and Deyerlien. *(Rees)*

One of the distinguishing features of the pioneers is that many were migrants to the London area coming from as far afield as France, Germany and Scotland, with Maudslay himself a noted exception. Once settled in London, most of these men remained and passed on their businesses to sons and sometimes grandsons, establishing engineering dynasties. Both Hall's and Donkin's firms have now lasted for nearly two hundred years. Many of the engineers already noted played a leading role in attempts to organise and institutionalise the engineering profession. The idea of a federation of engineering employers was first given practical shape in 1805 with the foundation of the short-lived Society of Millwrights. Among its members were Donkin, Rennie, Penn and Hall. Later, in 1818, Donkin together with two of Maudslay's sons, and his partner, Joshua Field, was instrumental in founding the Institution of Civil Engineers. But no matter how able or highly regarded the master engineer might have been, he remained a general mechanical engineer throughout this period. Hence virtually all the engineering manufacturers examined so far accepted orders for a wide variety of products often to customer specification. There was insufficient demand to support specialist engineering firms though certain firms could specialise in the short term to satisfy a particular bulk order.[29] This was despite the additional engineering demands resulting from the Napoleonic Wars of 1803 to 1815. During this conflict Maudslay's rise to fame was associated with his construction of Brunel's block-making machinery but he also supplied a number of steam engines to the Royal Arsenal together with machinery for the manufacture of Congreve's rockets, while Bramah supplied a hydraulic machine for lifting guns and drawing piles.[30] Both John Hall and John Penn supplied machinery for gunpowder mills.

By the 1820s, Maudslay's influence would have extended to the leading machine shops where compound slide rests would have been common items and the power source might have been a simple A-frame beam engine or a compact table engine on the Maudslay principle.[31] These prime movers communicated power to the individual machine tools by a system of pulleys, belts and line shafts. There was a pyramidal pattern of organisation among the engineering businesses, with a small number of large and mature firms, a modest number

Fig.21 Bramah's hydraulic machine for drawing piles. *(Rees)*

of medium-sized workshops and a large number of small concerns. The large-scale factories would have had their own pattern shops, iron and brass foundries and smiths departments, while the smaller businesses concentrated on machine shop work, ordering castings and forgings from outside suppliers. In 1825, according to Alexander Galloway, there were 'not many leading shops of great magnitude' but he knew of 'half a dozen masters who are employing now 30 or 40 men each'. At this time there were between 400 and 500 different masters working in the greater London area in various branches of machinery and between 9,000 and 10,000 men connected with the machine trade.[32] If these figures are accepted as accurate[33] then many of the master engineers would have been employing fewer than ten men. Undoubtedly one of the largest firms was Maudslay's with around two hundred men[34] and other major enterprises included Galloways with over 100, Taylor and Martineau, and many of the firms already mentioned such as Bramah & Sons, G&J Rennie, John Penn & Sons and John Hall & Sons. The major firms would have centred their activities around steam engine manufacture with its new applications to ships and other forms of transport such as steam road cars and coaches.[35] John Braithwaite junior (1797-1870), who carried on his father's New Road engineering factory, became acquainted with Capt. John Ericsson in the late 1820s which led to Messrs Braithwaite and Ericsson constructing the *Novelty* for the Rainhill locomotive trials. This engine was the first to complete a mile within the minute.

A new entry to the industry, which had a Maudslay connection, was the firm of Messrs Seaward & Co.[36] The Limehouse Iron Works, Millwall, was established by John Seaward (1786-1858) in 1824 and he was joined by his younger brother, Samuel (1800-1842) in the following year. The education, training and background of both men prior to their joint venture was both broad and varied, underlining the need for more than a single craft or type of work experience if success in the increasingly competitive engineering industry was to be achieved. After an education in classics and mathematics John Seaward first worked for his father, a successful builder, as an architect and surveyor. He then worked for several years for Messrs Doulle, government contractors, as surveyor and works manager. Further management experience was gained in Wales in charge of a number of lead mines. Here he met Woolf and Richard Trevithick and acquired some knowledge of chemistry and mining machinery. On return to London he superintended the construction of several private docks and became agent for a Staffordshire iron works and the Imperial & Continental Gas Co. providing gas lighting for a number of towns and cities in northern Europe. Before moving to Millwall he had submitted a design for a new London Bridge with three arches of 230ft span. Upon setting up in business, therefore, John Seaward was an accomplished civil engineer with considerable management and commercial acumen and knowledge of mechanics. The younger brother Samuel Seaward first entered the service of the East India Company as a midshipman. After two voyages to India and China he returned to London and took up an apprenticeship with Maudslay. After five years at the Lambeth works he joined Taylor and Martineau and worked in Cornwall under Woolf erecting several large pumping engines. His last post before joining his brother was superintending part of the Hayle Foundry where he worked with Richard Trevithick. Messrs Seaward & Co. were general engineers and designed and built large swing bridges, hoisting shears, dredging machines, cranes and equipment for docks and mills of various sorts. The firm provided the swing bridge for Telford's St Katherine's Dock in 1827. The enduring reputation of the firm, however, was made through its proficiency in the design and construction of marine engines. In 1829 Seaward's helped form the Diamond Steam Packet Co., building many of the boats which ran between London and Gravesend. Both brothers became members of the Institution of Civil Engineers.

After visiting Taylor and Martineau's extensive works in the City Road and admiring 'such large and fine products leaving their plant' the Swiss traveller J.C. Fischer (1773-1854) lamented that he had spent the best years of his life in a country where the production opportunities were so restricted.[37] Both Philip Taylor and John Martineau came from well-established Norwich families linked by long-standing friendships, unitarianism and marriage ties.[38] The Martineau family was probably the more prosperous, with business interests in brewing, banking and sugar refining together with shares in Wheal Friendship, a large Devonshire copper mine. Philip Taylor (1786-1870) first studied surgery but then became a druggist, inventing wooden pillboxes and making the first specimens by a lathe allegedly turned by a pet spit-dog. He moved to London in 1815 to be a partner in the chemical works of his brother, John. In 1819 John Martineau joined the Taylor brothers and in 1820 they expanded the business into mechanical engineering, manufacturing portable printing machines for Marc Brunel. Martineau was clearly an engineer of repute for he was called as first witness in 1824 and 1825 before the Select Committees of the House of Commons investigating laws prohibiting the emigration of skilled workers and the exportation of machinery. He stated before the latter committee that 'the manufacture of steam engines, and the manufacture of apparatus for making gas from oil, form a considerable part of our business'.[39] In 1824 he reported that the main component of the firm's exports was 'apparatus for the making of oil gas into Paris'. Despite the size and reputation of Taylor and Martineau, little is known about its activities beyond the observations recorded in Fischer's diary. When Fischer visited the firm's workshops in July 1825 he saw many machines being built including eighteen steam engines, several gas generators, gas cylinders, gas compressors and pumps. On another occasion he inspected the firm's laboratory where he saw an ice-making machine and a new type of power loom. Customers for Taylor and Martineau steam engines included Marc Brunel, the Thames Tunnel Co. and the London Portable Gas Co. Fischer visited the Apothecaries Hall, where he saw Taylor and Martineau's apparatus for generating gas from fish oil. This generator cost £170 and could make 80 cu.ft of gas out of a gallon of oil in an hour. Fischer praised the machine's 'simplicity and utility'.[40]

Another large firm whose very existence is apparent only through trade directories, House of Commons reports and the brief but disparaging remarks of Smiles, is that of Alexander Galloway. Joseph Clement said of him that he was 'only a mouthing common-council man, the height of whose ambition was to be an alderman'.[41] According to Smiles he 'very rarely went into his workshops to superintend or direct his workmen, leaving this to his foreman – a sufficient indication of the causes of his failure as a mechanic'. Apparently when Clement was first employed in Galloway's shop he was set to work at the lathe but found the tools so unsatisfactory that he went to the forge and made a new set for himself. Galloway was unprepared to recognise Clement's initiative or superior working skills in the form of a wage increase[42] and this caused Clement to leave for Bramah's factory within a matter of weeks. To compound Smiles' indictment of Galloway, he relates how the engineer constructed a cast iron roof for his workshop 'so flat and so independent of ties that the wonder was that it should have stood an hour'. When the engineer Peter Keir was shown the new construction he was so alarmed by the 'most unprincipled roof' that he insisted on continuing his conversation with Galloway outside the premises. The roof collapsed the following morning killing ten men.[43]

Smiles most probably obtained his information on Galloway from Clement's nephew, Joseph Wilkinson, and from John Penn junior (1805-1878).[44] It is possible that personal animosity, business rivalry or professional jealousy affected the assessment.[45] No mention was made of the fact that Maudslay's roof also collapsed in 1826, earning an anonymous rebuke from George Rennie.[46] Galloway began his engineering enterprise in 1801.[47] Between 1812 and 1824 he employed a cumulative total of between 1,000 and 1,500 men. In 1824 his workforce numbered about 80 but this had grown to between 110 and 120 by the following year. Despite his criticisms, Smiles admitted that Galloway was 'one of the leading tradesmen of his time'. In the early 1820s Galloway invested some £30,000 in a new factory. He accepted orders for steam engines, hydraulic presses, minting machinery, saw mills, iron and copper mills, and was prevented from manufacturing textile machinery – Galloway was obliged to refuse two orders worth £15,000 each for bobbin net machinery – for the continent only by the laws prohibiting the export of this type of machinery. Galloway's exported to France, Russia, Egypt and Algiers and was, according to its founder, the capital's principal exporter to South America. Galloway described himself in 1817 as a 'machinist and engineer, residing in Holborn'.[48] From his own account it would appear that he was an extremely enlightened employer: no man was employed at his works unless he could read and write and produce a good character; he did not permit 'obscene and vulgar language' or fighting in the manufactory. Transgressors were fined by the men themselves; these fines were paid into a sick fund into which all employees above labourers paid 6d per week;[49] and if a man were off work through sickness then he received a benefit of £1 per week. Galloway

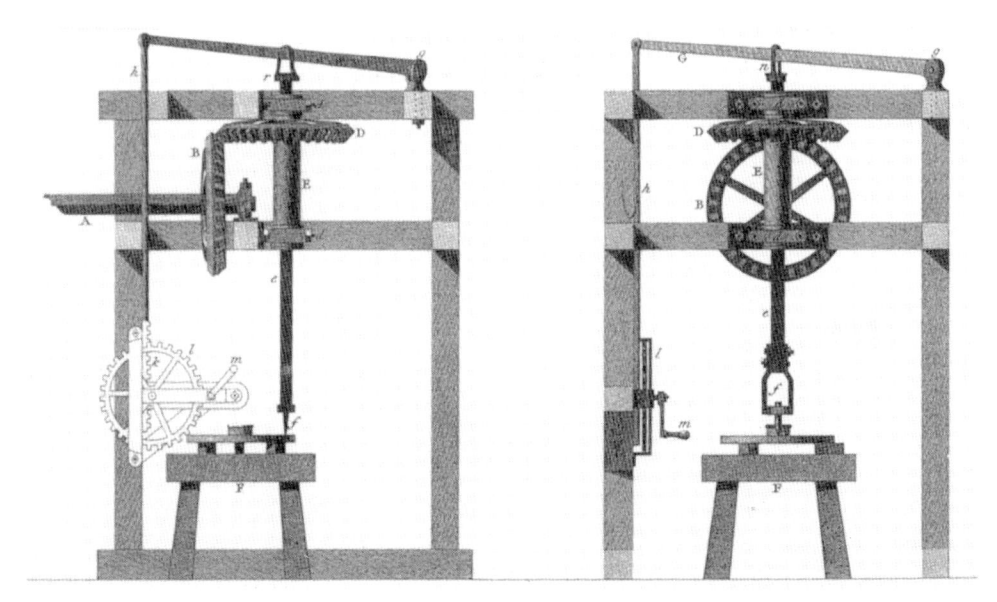

Fig.22 Early nineteenth-century workshop drilling machines. *(Rees)*

referred to the latter system as a 'cheap benefit fund' and claimed that similar systems were operated by the larger metropolitan engineering employers. By these means Galloway avoided labour disputes and instilled such qualities as good conduct, deportment and regularity among his workforce.

The first generation of mechanical engineers, of whom Maudslay was a leading member, provided training grounds for the second generation who, unlike their predecessors, often chose to pursue their engineering careers away from the capital. After assimilating the lessons of their masters, Fairbairn, Roberts, Lewis, Whitworth, the Nasmyths and Muir all headed for Manchester where the continuing expansion of the textile industry and the birth of the railway age offered unprecedented opportunities to young engineers with talent and enterprise.[50] The North West also provided more space, lower rents and cheaper labour. But this did not mean that London ceased to be a major centre of the engineering industry. By 1851 the capital still contained more than 6,500 engine- and machine-makers and was outnumbered only by Lancashire and Cheshire, with over 14,000 of the same. London's continuing importance owed much to the growth of marine engineering, a development in which Maudslay's played a leading role. By the 1850s there were some 1,200 men employed at the Lambeth factory.[51] While only Clement and David Napier, among Maudslay's more eminent alumni, chose to remain in London, a number of the Manchester engineers returned at a later stage in their careers. In doing so they were either hoping to engage in the shipbuilding

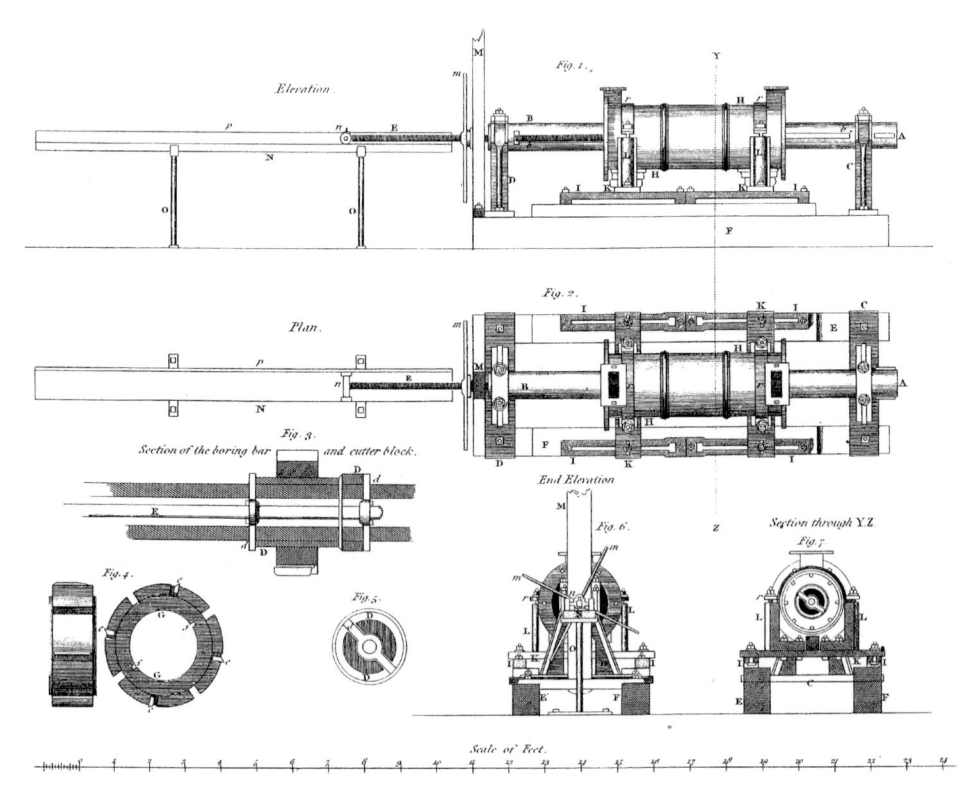

Fig.23 Early nineteenth-century workshop cylinder boring machine. *(Rees)*

boom, as was the case of Fairbairn,[52] or to practise as consulting engineers, as with Roberts and George Nasmyth. London's role in the development of the early mechanical engineering industry is almost impossible to overstate, for although there were individual firms of excellence located in the provinces during the critical formative years,[53] the greatest concentration of engineering skill and the greatest opportunity for the exercise of that skill remained in the nation's capital. It was because of this that the leading firms, especially Maudslay's, became 'a Mecca for aspiring young engineers and a seed-bed of engineering talent'.[54]

Endnotes

[1] There is no full-scale study of the London engineering industry at this time, though there are a number of short surveys of individual firms and some of the more prominent engineers. References are made to these works in the course of this chapter.

[2] Shaping, slotting and key-grooving machines did not appear in the machine shops until the 1830s.

[3] The West India Docks were opened in 1802, followed by the London Docks (1805) and the East India Docks (1806).

[4] *S.C. on the Exportation of Tools and Machinery*, P.P., Vol. 5, 1825, p.134. Clock and watchmakers converting to mechanical engineering were more likely to work as journeymen than entrepreneurs.

[5] In West Yorkshire most of the early engineers were former smiths rather than clockmakers or millwrights: Cookson 1996.

[6] Musson & Robinson 1969, pp.372-92.

[7] The drop stamp was used in the batch production of small metal goods such as button fronts.

[8] Each of these dockyards would have had its smithy where anchors and other metal parts were produced. The equipment of the smiths included a hearth, anvil and range of hammers, tongs and swages. Steam-powered tilt or helve hammers were used for large forgings, and the foot-powered Oliver, a light hammer requiring a single operator, for the production of small parts such as bolts or rivets.

[9] Hogg 1959-60 and Rolt 1965, pp.45-48.

[10] Wooden-bed lathes could alter dimension with changes in temperature and humidity.

[11] Smiles 1863, p.180.

[12] According to John Farey, by 1805 some 112 steam engines were at work in London: twenty-five for pumping water for public supply, seventeen in breweries, ten in foundries and machine makers, and the rest in a large range of trades ranging from flour mills to diamond cutters and silversmiths: Farey 1827, p.654.

[13] According to Gregory the 'large lathes' in Maudslay's workshop were worked by hand; the men turned a fly-wheel which was belted to another wheel 'fixed to the ceiling, directly over the mandril of the lathe'; the latter wheel worked the mandril 'by a band of catgut': Gregory 1806, p.474.

[14] Dickinson 1942 and McNeil 1968.

[15] According to *The Post Office Directory* for 1800, Bramah had premises in both Piccadilly and Pimlico at that date. This is not acknowledged by McNeil 1968.

[16] In modern day terms a model shop would have been a Research and Development Department.

[17] According to J.C. Dyer, Fox of Derby saw Bramah's wood planer and adapted it for metal working about ten years before Richard Roberts made his machine: Dyer 1868, p.148.

[18] Boucher 1963.

[19] Rennie made a verbal agreement with James Watt not to manufacture steam engines in London: Boucher 1963, p.10.

[20] Musson and Robinson 1969, pp.63 & 78.

[21] Hesketh 1935.

[22] Hall appears several times in the travel diaries of Joshua Gilpin, the American papermaker, who toured the mills of Kent and Hertfordshire at the end of the eighteenth century: Woolrich 1996 & 1997.

[23] Donkin 1949-51.

[24] Barry 1863, p.267, and *DNB*.

[25] Woolf came to London from Cornwall in the late 1780s and entered the service of Bramah before taking up the position of resident engineer at Meux's brewery. Between 1806 and 1812 he was a partner in a Lambeth steam engine factory with an engineer named Edwards. Woolf took out a number of patents in connection with steam engine design and made important improvements to machine tools for steam engine production, including the introduction of all metal lathes. He also introduced more accurate machining methods in the finishing of castings. In 1812 Woolf returned to Cornwall where, until 1833, he acted as superintendent of Harvey & Co.'s engine manufactory at Hayle, where he undertook the introduction of London manufacturing machines and methods. During this time William Muir (see Chapter 9) undertook part of his training at Harveys: Harris 1966.

[26] Braithwaite established an engineering works on the New Road, London, in the late eighteenth century where he devised and manufactured machinery and equipment in connection with his salvaging activities, including diving apparatus and machinery for sawing ships asunder under water. After 1818 his business was continued by his two sons, Francis and John.

[27] Dixons of Maid Lane, Southwark, were manufacturers of steam engines and iron printing presses.

[28] Dyer 1868, p.141. Presumably the capital resources of this firm would have been supplied by Sir John Henry Pelly (1777-1852) who was a governor of the Hudson Bay Co., a member of Trinity House and a governor of the Bank of England: *DNB*.

[29] Maudslay told a select committee in 1825 how his firm had dedicated its activities to the production of ship's tanks for six months with an output of ninety-eight per week: P.P., 1825, Vol. 5, p.32.

[30] Hogg 1959-60.

[31] See pp. 25-26.

[32] P.P., 1825, Vol. 5, p.156.

[33] Some doubt is cast on Galloway's figures by the fact that in the previous year he had claimed that there were between 200 and 300 master engineers in the metropolis and its vicinity: P.P., 1824, Vol. 5, p.19.

[34] See p.19.

[35] Manufacturers of these vehicles included Walter Hancock, Sir Goldworthy Gurney and Bramah who made the steam carriage for Julius Griffiths. The machines were developed with a view to running fare-paying services. Companies such as the London and Birmingham Steam Carriage Co. and the London, Holyhead and Liverpool Steam, Coach and Road Co. were floated but quickly foundered. Technical problems, opposition from vested interest groups, such as the various London omnibus companies and stage coach proprietors, together with managerial incompetence, ensured the death knell of this early experiment in mechanical road transport: Nicholson 1982.

[36] *DNB* and Chrimes et al 2002.

[37] Henderson 1966, pp.27-29.
[38] Burt 1977, pp.15 & 22.
[39] P.P., 1825, Vol. 5, p.138.
[40] Henderson 1966, p.35.
[41] Smiles 1863, p.241.
[42] Galloway claimed that he refused to operate a flat wage rate system and negotiated individual wage contracts with each of his workmen: P.P., 1824, Vol. 5, pp.27-28.
[43] According to *The Times*, 6 September 1824, there were only two fatalities though another man was 'not out of danger' in hospital and several men were injured.
[44] All the information about Galloway in Smiles 1863 is contained in his chapter on Joseph Clement. For John Penn junior see Cookson and Hempstead 2000, pp.26, 99, 100.
[45] Galloway appeared as a witness before House of Commons Select Committees in 1817, 1824 and 1825.
[46] Petree 1934, p.586.
[47] The following information is taken from Galloway's evidence to a 1824 Select Committee: P.P., 1824, Vol. 5, pp.14-28.
[48] P.P., 1817, Vol. 6, p.246. Galloway makes his first appearance in the *Post Office Directory* in 1806 with premises at 69 High Holborn.
[49] Labourers paid 3*d* per week to receive a sick benefit of 10*s* per week.
[50] John Hall trained the Bertram brothers, William and James, who later established a paper-making works in Edinburgh.
[51] Petree 1934-35, p.43.
[52] William Fairbairn (1789-1874) opened his shipbuilding works on the Thames at Millwall in 1835 while still retaining his Manchester works. The former turned out to be a financial disaster and was eventually disposed of in 1849.
[53] Apart from Boulton & Watt's Soho factory near Birmingham the most noted engineering enterprises outside London at this time were those established by James Fox (*c*.1760-1835) of Derby and Matthew Murray (1763-1826) of Leeds. See Smiles 1863, pp.258-59 and 260-64; Kilburn Scott 1928; Tyas 1925-26; Forward 1937-38; Turner 1966. Murray and Fox were among the first to manufacture machine tools for sale, both at home and abroad, early in the nineteenth century. Examples of Fox's machine tools may be seen in the Birmingham Museum of Science and Industry.
[54] Musson 1978, p.116.

4
Richard Roberts
Richard L. Hills

It was said of Richard Roberts that 'although our records are crowded with the names of eminent engineers, he is the greatest mechanical inventor of the nineteenth century'.[1] Richard Roberts might be called the father of modern production engineering. He worked for Henry Maudslay between 1814 and 1816, after which he returned to Manchester to establish himself in business on his own. The influence of Maudslay was probably critical for changing Roberts from working in wood to working in metal and so laid the foundations for his later career. Throughout the rest of his life, Roberts used his inventive genius both to improve a wide range of existing machines, most notably the power loom and the spinning mule, and to develop machine tools to produce his inventions. One invention stimulated another, and this combination of more accurate and specialised machine tools being used to fabricate better production machines led to standardisation of parts, and hence an early form of mass production which in turn demanded more specialised tools.

Roberts' father, William, a shoemaker, kept the New Bridge tollgate at Llanymynech on the border between England and Wales, with the front door opening into one country and the back into another. Although Roberts could have been born in either country, he always considered himself a Welshman. He was born on 22 April 1789, the second of four sons and there were also three younger daughters. Such education as Roberts received came through the local parish priest, the Revd Griffith Howell, who ran a school in the church belfry. Howell found the ten-year-old Roberts carving the handle of a walking-stick when he should have been doing his lessons, but instead of reproving the lad, he encouraged the boy to use his lathe and other tools. Roberts made a spinning wheel for his mother and later said that he 'taught himself wood turning, in which art he always thought he excelled any other man'.[2]

Roberts found early employment with a boatman on the Ellesmere Canal and then at the local limestone quarries. He continued with his woodworking activities so that his fellow workmen clubbed together to buy him a set of tools. He was able to repair a lathe and received instruction in drawing from Robert Baugh, a road surveyor. In 1851, Roberts told the Select Committee of the House of Lords for the Patent Law Amendment Act that 'I make my own drawings, and my draughtsman copies them afterwards, and fills in the details'.[3]

At the age of twenty, Roberts found employment as a pattern-maker at Bradley Iron Works, Staffordshire, where he may have met Thomas Jones Wilkinson, nephew of John Wilkinson, and Roberts' future partner in various companies in Manchester. After gaining further experience elsewhere, Roberts became a leading hand in the pattern shop of the Horseley Iron Works, Tipton, probably in 1813. He was regarded as 'a sort of jack-of-all-trades, for he was a good turner, a tolerable wheelwright and could repair mill-work at a pinch'.[4] In order to escape being recruited into the militia, Roberts left Horseley, said farewell to his father at Llanymynech and made for Liverpool. Finding no work there, he went on to Manchester where he was able to help a cabinet-maker turn up some bedposts. He obtained

Fig.24 Richard Roberts. *(Illustrated London News, 1864)*

the post of chief turner with this person before finding another job in Salford, lathe- and tool-making.

Because the militia officers were still seeking him, Roberts decided to seek refuge in the anonymity of London and walked there in the company of two fellow engineers, Francis Lewis and Murgatroyd.[5] Both later worked for Roberts on his return to Manchester and Lewis eventually established his own factory for machine tools. Roberts succeeded in obtaining employment with Maudslay as a turner and fitter. There he imbibed the philosophy of Maudslay, regarding the importance of accurate machine tools where hand-work was replaced by mechanisms. While in London, Roberts strove hard to augment his rudimentary education.[6]

Roberts worked at Maudslay's for some time, 'acquiring much valuable knowledge in the use of tools, cultivating his skill by contact with first-class workmen, and benefiting by the spirit of active contrivance which pervaded the Maudslay shops'. Having greatly improved his manual dexterity and stimulated his 'inventive ingenuity', he decided to return to Manchester, which offered 'openings for men of mechanical skill' even more abundant than London itself.[7]

The defeat of Napoleon in 1815 removed the threat of being called up for the militia, so it was safe for Roberts to return north. By 1816, he had set up in Manchester as a 'turner of plain and eccentric work' at No.15 Deansgate.[8] The lathe was upstairs in a bedroom, driven from a big wheel in the basement by his wife.[9] We know nothing else about this woman, neither her name, marriage nor death. He soon moved into the New Market Buildings at Pool Fold where he was described as 'Lathe and Tool Maker'.

Fig.25 Front view of Roberts' gear-cutting machine with some teeth already cut on the vertically mounted blank. *(R.L. Hills)*

Roberts quickly constructed a range of machine tools, some of which were to his own designs, the first being a gear-cutting machine. His later machines of this type had the wheel to be cut, which could be over 3ft in diameter, mounted vertically. A rotary cutter sliding underneath the gear blank cut one tooth at a time after which the blank was indexed round another space, following the principle of machines used by clock and watch makers. For accurately checking the outer diameter of the blank, Roberts adapted a measuring instrument, the sector, which was developed for sale. Roberts had seen rotary cutters being used at Maudslay's and later extended their role in his own workshops.

In 1817, Roberts built a lathe capable of turning work up to six feet long, having centres 9½in high. It featured backgearing to give an increased range of spindle speeds, a device probably invented by Roberts. He had learned from Maudslay the principle of holding the cutting tool in a slide rest to give greater accuracy in turning than with hand-held tools, and on this lathe Roberts fitted a sliding saddle which could be moved through bevel gearing as the work rotated. When the saddle had travelled far enough, it could automatically knock itself out of gear. This early example of a true industrial lathe can now be seen in the Science Museum, along with Roberts' gear-cutting machine.

Roberts used the slide principle again in 1817 when he built his first planing machine, now also at the Science Museum. A man employed to chip and file flat surfaces spent his wages on drink, with the result that Roberts became tired of the irregularity of this person's work and decided to make a machine to replace him. The result was the planing machine, always considered by Roberts as of his origination, although this honour has been claimed for others. The table on his first machine was 52in long by 11in wide. Although people were allowed to visit Roberts' works freely and so could have seen this and other machines, it was

Fig.26 Roberts' back-geared lathe, c.1817. *(Science Museum/SSPL)*

not until 1825 or 1826 that one was sold to Maudslay and it was some years later that this versatile machine became common in other engineering establishments.

Roberts had imbibed the importance of machine tools from Maudslay so that he was filling his own workshop with other types. An advertisement in 1821 mentioned screw-cutting machines and also indicated that Roberts could divide quadrants as well as cut gears.[10] When Joshua Field visited in August 1821, he saw at least four lathes of different sizes, screw and gear-cutting engines, a drilling machine and two forges. Twelve or fourteen men were employed.[11] There are no details of how Roberts financed all this for he could not have saved much while he was in London and there is no indication that he had formed any partnerships up to this time.

Apart from letter-copying presses, we have a little knowledge about other products from this workshop. In 1817 or 1818, Roberts was asked by the Police Commissioners of Manchester to develop a meter to measure the supply of gas to their customers. Using a water seal, he succeeded where Samuel Clegg in London had failed. Roberts could not afford to patent his invention which was quickly copied by Clegg. In 1818, Roberts made a 'rifled brass cannon adapted to load at the breech with which Mr Bradbury experimented on

Fig.27 Roberts' original planing machine of 1817. *(Science Museum/SSPL)*

> RICHARD ROBERTS,
> *Lathe, Screw, Screw-Engine, Screw Stock, &c., &c.,*
> *Manufacturer,*
>
> RESPECTFULLY informs COTTON-SPINNERS, IRON-FOUNDERS, MACHINE-MAKERS, and MECHANICS in general, that he has CUTTING-ENGINES at work on his NEW and IMPROVED principle, which are so constructed as to be capable of producing ANY number of Teeth required: they will cut BEVIL, SPUR, or WORM Geer, of any *size* and *pitch* not exceeding 30 inches diameter, in WOOD, BRASS, CAST-IRON, WROUGHT-IRON, or STEEL, and the TEETH will NOT REQUIRE FILEING UP; DIVISION-PLATES, QUADRANTS, &c., accurately divided, or additional Numbers put on *Old Plates.*
>
> N.B. R. R. cuts, on his *Improved Screw Engine*, SCREWS of ALL Sorts, Pitches, or sizes, with the greatest accuracy.
>
> Manufactory, New Market Buildings, Pool Fold; House, 5, Water-street, Manchester.

Fig.28 The advertisement which Roberts placed in the first issue of the *Manchester Guardian*, 5 May 1821.

spherical missiles coated with lead'.[12] This was well in advance of other guns at that time since it was only during the Crimean War that rifling and breech loading became common.

During the latter part of 1821, Roberts moved his business to the Globe Works in Faulkner Street where it remained until closure in 1852. This may reflect greater financial backing through association with four people: Thomas Sharp, an iron merchant and former agent for John Wilkinson; Robert Chapman Sharp, his brother; Thomas Jones Wilkinson; and James Hill. This group had acquired the rights to the British patent taken out by an American, Jeptha Avery Wilkinson, for a machine to make reeds for looms. Roberts was asked to improve this and Field saw his early attempts. Roberts' alterations were so successful that a company, Sharp, Hill & Co., made reeds for many years at Faulkner Street using Roberts' machines. Patent rights and machines were later sold in France.

The reed-making machines may have been the means of introducing Roberts to textile machinery. He must have seen hand-looms at work, which led him into the first of his twenty-five patents, that for a power loom, in 1822.[13] The date of the patent application in November almost coincides with Hill dissolving one partnership and entering another. 'Roberts, Hill & Co., Machinists and Engineers, Falkner Street' advertised at that date for turners and filers.[14] Presumably Hill financed the patent.

Roberts took out this patent a little after William Fairbairn had started replacing earlier cast iron line shafting in textile mills with lighter, stronger wrought iron. Not only did this increase the speed but also the power that could be transmitted, with the result that larger mills could be built. It may well be that Roberts was able to take advantage of this with his more accurately constructed looms made from iron which could withstand better the shocks

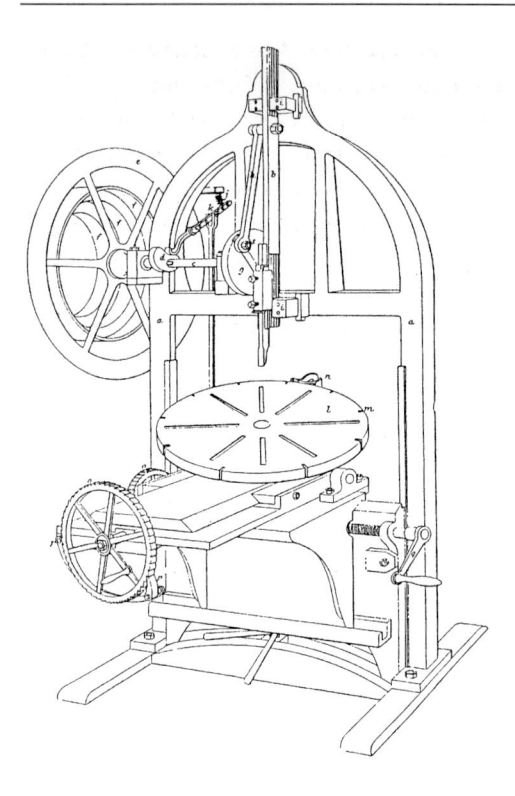

Fig.29 Roberts' slotting machine. *(Buchanan, 1841)*

of being driven at higher speeds. These looms were described as 'the real starting point of modern power loom weaving'.[15] J.C. Fischer of Schaffhausen was shown round the workshops of Sharp, Hill & Co. in June 1825 by Roberts. Fischer estimated that power looms were being turned out at the rate of over eighty a week, or roughly 4,000 a year, such was their popularity.[16]

Although this figure would not be considered mass production today, it must have stretched the capacity of the existing machine tools and workforce. Gearwheels and pulleys had to be secured to their shafts by keys driven into slots. To replace hand chiselling and filing the slots, Roberts again turned to the slide principle and invented the key-grooving machine in 1824. It has been said that this was based on the mortising machine made by Maudslay for Brunel's blockmaking machinery at Portsmouth. Here a crank drove an arm reciprocating vertically, with a cutting chisel secured to the end. The wheel to be slotted was mounted on a table which could be moved backwards or forwards to deepen the slot for the key. Later this machine was developed into the more versatile slotting machine on which the table could be moved sideways, rotated, and even tilted as well. Then straight or curved slots could be cut, edges trimmed and keyways cut in tapered holes. Another machine which Roberts derived from the key-grooving machine was the shaping machine in which the cutting tool was moved horizontally over the workpiece, rather like filing, while the workpiece could be moved vertically, sideways, and rotated, so that a great variety of shapes could be machined.

In 1823, Roberts was elected a member of the Manchester Literary and Philosophical Society, the leading scientific society in the city, where he met other prominent engineers.

He was a member of Council between 1852 and 1859 and elected an honorary member in 1861. With the help of other members, he carried out experiments which would then have been termed scientific. In October 1824, while installing what may have been his own design of blowing apparatus for the foundry at the Globe Works, he discovered what became known as the 'disc phenomenon'. The jet of air hitting the base of the valve diverges to escape, forming an area of low pressure in the centre, sufficient to retain the valve or disc a small distance from the orifice. These experiments were later described in a lecture to the society by Thomas Hopkins, assisted by Roberts.[17] A little later, Roberts built a rig on which a model of a railway wagon was mounted above a drum which could be rotated at known speeds to determine whether friction increased.[18]

Roberts was a leading figure in launching the Manchester Mechanics' Institution, now the University of Manchester Institute of Science and Technology. At a meeting in April 1824, Roberts seconded the resolution establishing the Institution and was a member of the organising committee. It opened in the following March with 400 members. Roberts must have recognised his own limitations through lack of formal education and wished that other artisans should have the chance to improve theirs. He remained connected with the Institution for many years, serving on various committees as well as providing demonstration apparatus.

It was in 1824 that Roberts conceived the idea for his most famous invention, the self-acting spinning mule. By his own account, the mule was developed after spinners in Hyde had been on strike for three months. A 'deputation of masters' called, and asked him to consider making the mule self-acting. 'I said that I knew nothing of spinning, and therefore declined it; they called a second time… I declined again; but before seeing me on the third Tuesday, they saw my partner, the late Mr. Thomas Sharp, and requested that he would do what he could to induce me to turn my attention to it; on the third visit which they made, I promised to make the mule self-acting.'[19] One reason for Roberts' reluctance may have been that he did not wish to take work away from skilled men.[20] After four months, he had solved the problem and took out a patent in March 1825.[21]

In the spinning mule invented by Samuel Crompton in 1779, rollers drew out the cotton rovings and passed the fibres on to rotating spindles mounted on a moving carriage.[22] At the end of the draw, everything stopped. Then the spindles had to be turned backwards to unwind the yarn from the tip of the spindles down to where the cop was being built up. With the spindles rotating in the spinning direction again, the carriage was pushed back in by the spinner as he guided the yarn being wound onto the cop by a faller wire. When completed, the cop was unwound from the mule spindle and skewered on another in the shuttle, ready for weaving. It was necessary to wind the yarn on to the cop in a special way so that it could be pulled off during weaving without breaking or jamming. The spinner might have to put up the mule carriage over 2,000 times in the course of spinning one set of cops.

The initial spinning sequence was mechanised around 1792 but all attempts to make the winding-on automatic failed. Roberts introduced a camshaft which controlled the various stages and brought in the different mechanisms in four sequences. Then he guided the faller wire by an inclined plane which adjusted each draw to build up the cop. Finally, to compensate for the differing diameters of the cop at various stages of winding-on, he regulated the speed by the relative positions of faller and counterfaller wires. The first self-acting mule was completed on Saturday, 16 July 1825, the day before the Globe Works caught fire. Arson was suspected but never proved. A large part of the works, patterns and parts for power looms were destroyed but luckily not the sections with the reed-making machines, the mule or the foundry. The place was quickly rebuilt. Although insured, it cost the firm over £10,000 and men had to be laid off. Mule production had restarted by November and the first machines received glowing accounts, but in fact the method of winding-on did not prove to be satisfactory and needed modification.

Although few mules were made to this initial design, Roberts must have anticipated receiving many more orders since he made special templates to manufacture standard parts. He also introduced male and female standard plug and collar gauges for checking sizes of holes and bearings, later copied by Joseph Whitworth. The system of templates and gauges formed a great addition to the resources of mechanical engineering because they enabled identical parts to be manufactured. Roberts probably extended the system to cover power looms for this was the way he later built railway locomotives. About this time, he also introduced machines for making standard nuts and bolts, which a little later were turning out over a thousand of the ½in size each day.

To complete the story of the self-acting mule, Roberts took out a further patent in 1830.[23] He radically altered the headstock with all its control apparatus by combining mechanisms from both patents. He created the first true feed-back control system in the world by making the counterfaller wire regulate the rate of winding-on by altering the position of a nut on another mechanism he invented, the quadrant. The spun yarn was wound on through the movement of the carriage and the motion of the quadrant which related the speed of rotation of the spindles to the diameter of the cop. Woodcroft wrote in 1863 that Roberts' self-actor, 'generally admitted to have exceeded the most sanguine expectations of cotton-spinners', was still the mule most extensively used in cotton manufacture.[24] Roberts' camshaft, quadrant, and control of winding-on by counterfaller, remained virtually unchanged on the last cotton spinning mules built in 1927. His self-acting mule was Roberts' most spectacular achievement through the difficulties he had to overcome and the elegance of his solution.

Technically the improvements covered in the 1830 patent were a great success but at first sales were slow. Inevitably, others tried to copy the mule; in 1832 Sharp, Roberts & Co. settled out of court a case of patent infringement against Mr. Knowles of the Oxford Road Twist Co. Sales increased in the second part of the decade, but by 1839 only £7,000 had been recovered in profits against £12,000 expended on the 1830 patent. In that year, the Privy Council granted Roberts' plea to extend the 1825 patent for a further seven years 'on account of the great ingenuity and merits of the invention, and the obstacles that had from time to time been opposed to the patentee from deriving a fair remuneration for so important an invention during the original term of the patent'.[25]

On 31 May 1826 James Hill retired and the two earlier firms of Sharp, Hill & Co. and Roberts, Hill & Co. were replaced by Sharp, Roberts & Co.[26] The four remaining partners, Thomas Sharp, R.C. Sharp, Wilkinson and Roberts were joined by the Sharps' younger brother, John. Wilkinson and R.C. Sharp retired at the end of December 1836.[27] During 1826 and into 1827, Roberts paid at least three visits to Mulhouse in France because Sharp, Roberts had undertaken to set up a factory for Andre Koechlin to manufacture textile machines. Sharp, Roberts were providing the machine tools, and drawings and patterns for textile machines. The project was delayed through Roberts' serious illness in February and October 1827. The venture stirred up great hostility in Manchester through fear of continental rivalry, leading to loss of orders for power looms, and it does not appear to have been repeated. Surviving advertisements for machine tools at this time suggest that Sharp, Roberts were rather trying to expand this line of business. A few years later, the growing railway industry opened a new market for such products.

It might have been expected, with Roberts' keen interest in mechanical inventions and Thomas Sharp's position as one of the promoters of the Liverpool & Manchester Railway, that Sharp, Roberts would have become involved with steam locomotives much earlier than in fact they did. It was not until April 1832 that Roberts patented various inventions for road vehicles, some of which could be applied to railway engines.[28] This patent contains the only known instance of Roberts' involvement with stationary steam engines. His variable expansion gear with the inlet valve controlled by the governor was a very advanced concept and was not introduced on mill engines for many years to come.

The inventions in this patent for iron wheels, a steam brake, a differential, and a method for controlling the high water level in steam boilers, were almost certainly applied to Roberts' steam road carriage, first tested along Oxford Road on 9 December 1833. With fifty passengers, it must have weighed 11 tons, and although it reached 12mph it was deficient in steam-raising capacity. After modification, it was tried again in March 1834 when the boiler feed pumps did not work properly. They were found to be failing again on 4 April so a hurried return was made towards the Globe Works. However, before the carriage reached there, the boiler exploded, injuring some people and damaging neighbouring property. This was a bold experiment, for the carriage was designed to carry thirty-five passengers. It could have been Roberts' answer for road transport to compete with rail, but never ran again. His patent design of omnibus in 1856 failed to win a competition promoted by the London General

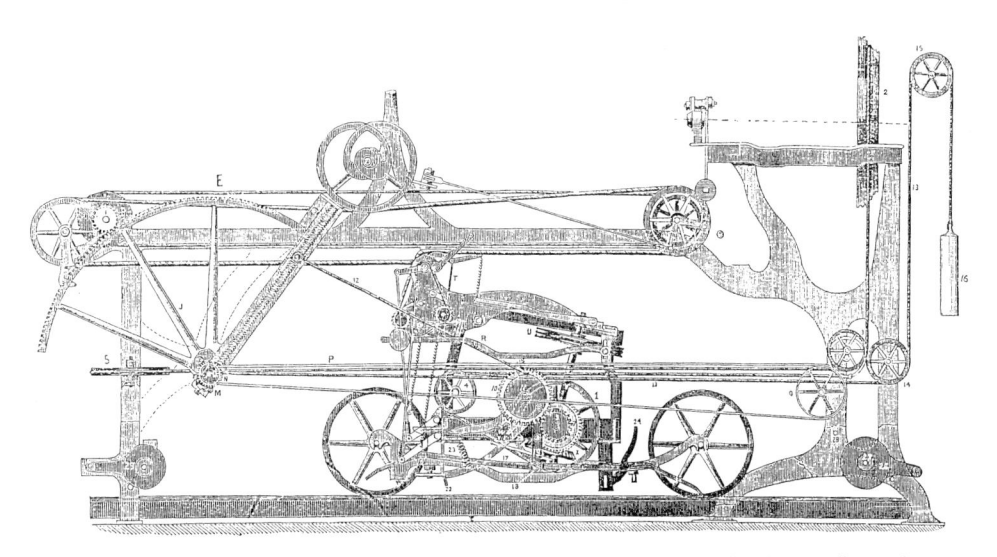

Fig.30 A cross-section of part of a self-acting mule headstock with the carriage for the spindles in the centre and the quadrant on the left. *(Leigh, 1875)*

Omnibus Company.[29] Around 1860, Roberts helped improve the Burrell-Boydell traction engine. These seem to have been the only links with road transport.

Nor was Roberts' initial foray into railway locomotive building successful. A 2-2-0 locomotive, the *Experiment*, was supplied to the Liverpool & Manchester Railway in May 1833. Roberts took note of some of Stephenson's improvements, such as the multi-tubular boiler and wheel arrangement, but redesigned the smokebox layout and placed the cylinders vertically between the wheels. The result was that this engine rode badly and consumed more fuel than comparable ones. In the following year, Sharp, Roberts supplied three more 2-2-0 to the Dublin & Kingstown Railway in Ireland with a conventional boiler layout. However, possibly in an attempt to improve the riding qualities, the cylinders were placed vertically over the front carrying wheels, making the engines pitch and roll even more violently. There were also problems with his design of cylindrical valves on all Roberts' early engines.

Whether it was perseverance that paid off, or whether it was through the introduction of the young German, C.F. Beyer, into the drawing office, the order from the Grand Junction Railway for ten 2-2-2 engines was received in November 1835, and marked the beginning of a fresh start in railway locomotive manufacturing. Sharp, Roberts quickly developed a more or less standard 2-2-2 design with curved outer frames over the driving wheels of usually 5ft 6in diameter, carrying wheels of 3ft or 3ft 6in diameter, inside cylinders with a bore of 12 or 13ins and an 18in stroke, and weighing around 12 tons. Orders for these engines were received in such numbers that the Globe Works did not have sufficient capacity. A new

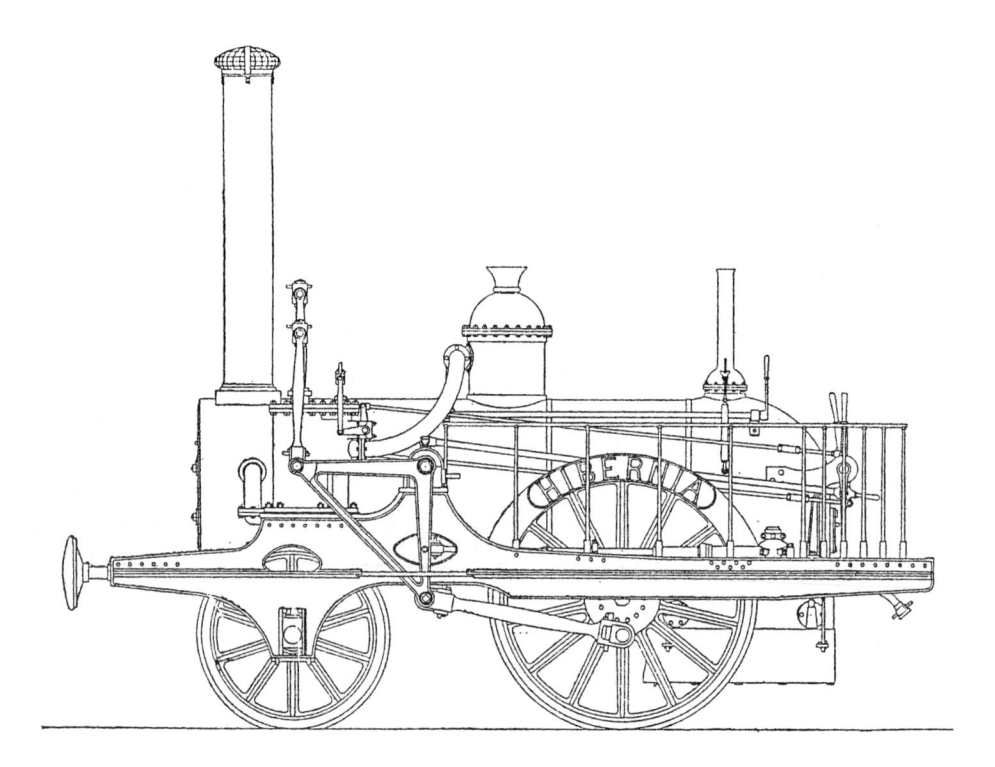

Fig.31 Hibernia sent to the Dublin and Kingston Railway, Ireland, in 1834. (Clark, 1860)

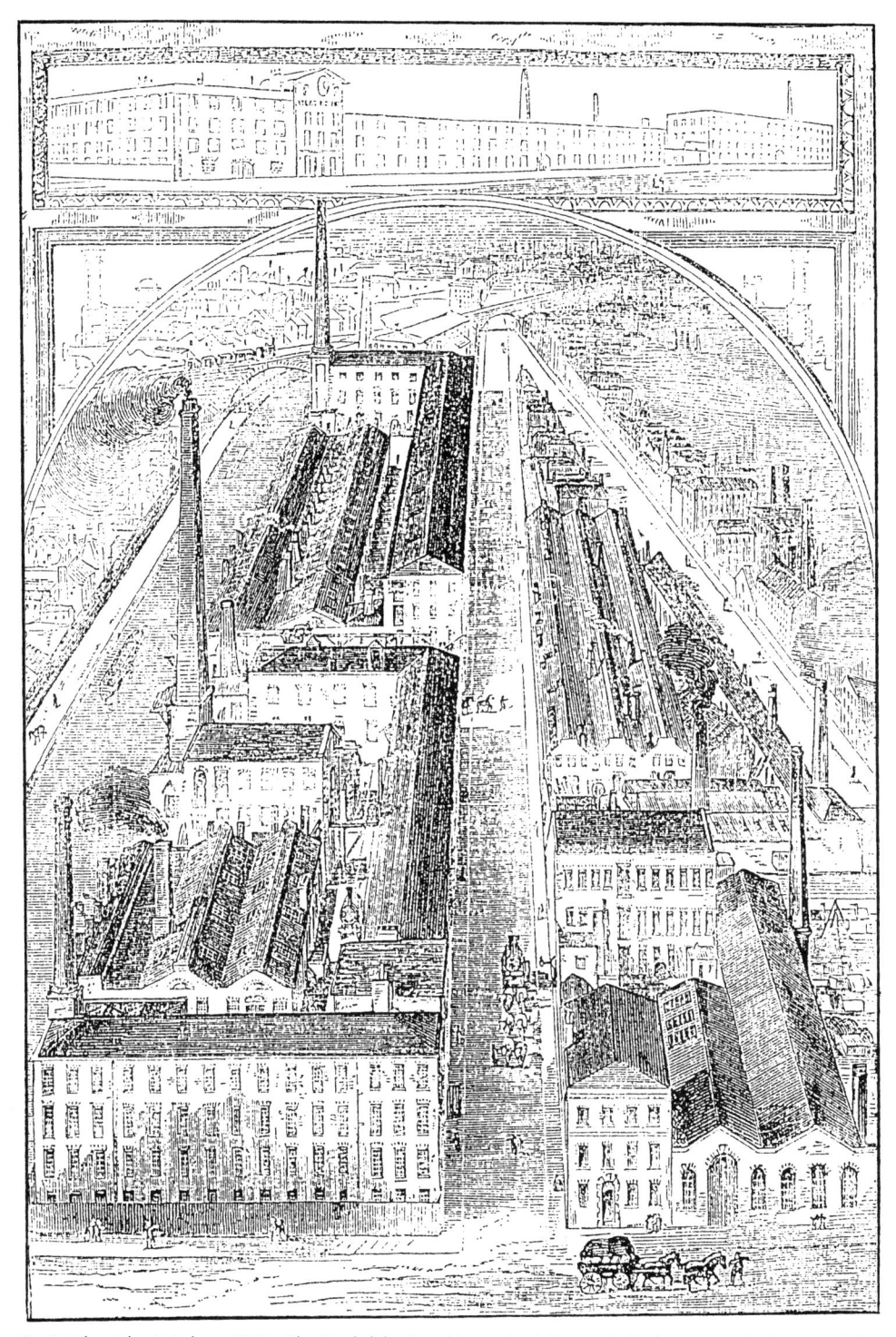

Fig.32 The Atlas Works, c.1854. The Rochdale Canal is on the left and Oxford Street at the bottom of the picture. Great Bridgewater street goes between the two sections of the works. *(Measom, c.1854)*

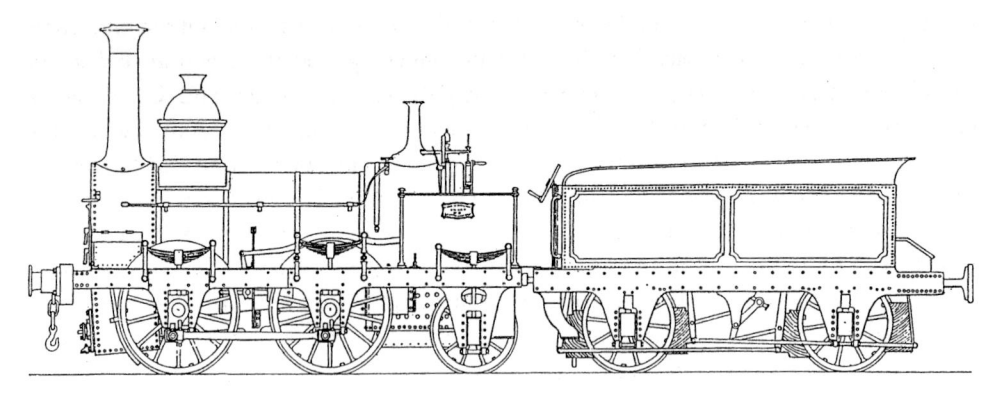

Fig.33 The Sharp, Roberts 0-4-2. *(Bennett, 1927)*

site, owned by Thomas Sharp, bounded by Oxford Road and the Rochdale Canal, was chosen for the Atlas Works, which cost around £30,000 and opened in 1839.[30]

Up to June 1843, Sharp, Roberts received seventy-nine orders for railway locomotives, resulting in the construction of around 246 engines. Of these, there were seven orders for fourteen engines for Ireland and forty orders for 111 engines for the continent.[31] The Sharp, Roberts engines soon attained a reputation for the excellence of their construction. Colburn wrote in 1871 of Roberts' 'immediate and important service' to locomotive construction. 'He introduced superior workmanship, giving better proportions and finish to the principal parts ... delighted in refined workmanship in connection with every kind of mechanism ... and was the first who brought to bear upon the modern locomotive that kind of craft which, while it did not alter its general mode of action, nevertheless contributed so much to its working economy.' Colburn identified in particular Roberts' stronger framing, better fastenings and larger bearing surfaces. 'In all parts machine-work was substituted to a great extent, in place of hand-fitting more commonly employed by the north country engineers ... Mr. Roberts added characteristic features of design, which still distinguish the English locomotive engine; and by his practice ... sensibly advanced the standard of locomotive construction upon every railway in the kingdom.'[32]

The influence of Maudslay can be traced directly in two ways in Roberts' improvements to locomotive manufacture. The first was in extending the use of rotary cutters. In 1834 Roberts introduced his slotting drill, evidently an early form of vertical milling machine. The drill cut grooves in bearing brasses for locomotive axles as well as bearings for other machine tools. A smaller version cut teeth on escapement wheels for his turret clocks.[33] William Davol, an American visitor in 1839, saw a large rotary cutter with a cast iron wheel of about 18in diameter and 3½-4in wide, with steel cutters around its periphery, machining locomotive crankshafts.[34] Having once discovered the efficacy of this method of metal removal, Roberts adapted the principle in other machines, such as one for forming the heads of bolts.

The second influence came through Roberts' knowledge of Maudslay's punching and shearing machines. Roberts was offering some such for sale in 1826 and used them later for boiler manufacture. Punching machines were used for cutting out the curved main frames for his locomotives. He improved the early type of shearing machine, resembling a giant pair of secateurs, by altering the angle of the blades so that the metal to be cut did not slip out. Then he introduced a more compact form in which the punching head at the bottom and shears above were worked directly by an eccentric. He may also have been responsible for a variation on this type in which the punching head was on one side and the shears on the other. This design was copied by other manufacturers and became very popular. In 1847, Roberts patented an improvement which changed the way in which the punch was engaged, and further adapted it in his last full patent in 1860 for punching holes in angle or T bars.[35]

Thomas Sharp and Roberts took out two patents in 1834. The first was for a corn-grinding machine in one version of which the stones might be set vertically. The second was for improvements to an American ring spinning frame.[36] The success of the power loom had increased demand for stronger warp yarn, spun most economically on the earlier throstle than the mule. The ring frame was to replace the throstle. It is probable that Sharp, Roberts wanted to be able to offer a broad range of textile machines because, in the following year, they began to manufacture another American machine for preparing rovings, the Eclipse Speeder. The ring frame continued to be made by Sharp Brothers, successors to Sharp, Roberts, who displayed one at the Great Exhibition in 1851.

In 1835, Roberts attended the British Association for the Advancement of Science meeting in Dublin, the first of many where he showed recent inventions such as a praxinoscope for viewing moving objects. In the late 1830s, Roberts started production of turret clocks, installing his first on the Atlas Works and another in the tower of Llanymynech church. At the same time he invented a centrifugal railway, its track in the form of a vertical loop-the-loop round which a truck could be sent. In a model, the truck was filled with water and none was spilt if sufficient speed were attained. Larger versions were made to carry a person in a Hall of Science in Manchester as well as pleasure gardens elsewhere.

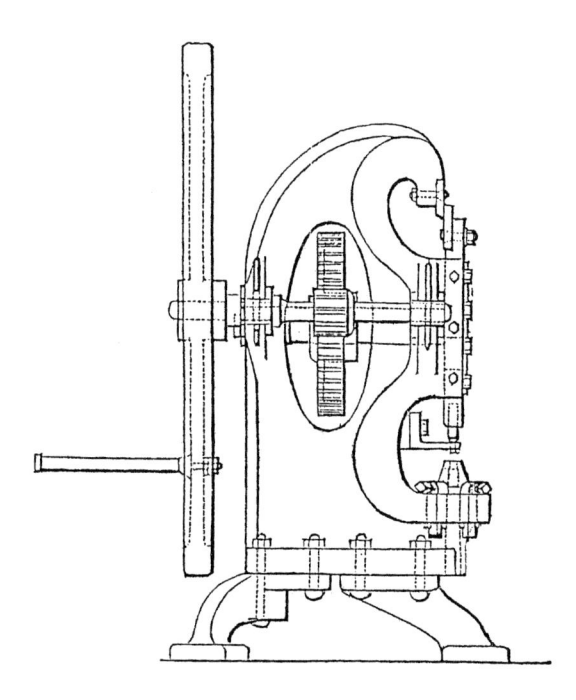

Fig.34 Roberts' small 'portable' punching and shearing machine with the punching head at the bottom and the shears at the top. *(Holtzapffel, 1846)*

On 20 March 1838, Roberts became a member of the Institution of Civil Engineers, the citation referring to scientific attainments as well as 'eminent practice as a civil engineer'.[37] In the same year, he was elected councillor for the Oxford Ward in Manchester, where his works were situated. He continued to represent this ward until 1843. Then in 1839 he was involved with launching the Royal Victoria Gallery for the Encouragement and Illustration of Practical Science, which was intended to stimulate research and invention with an impressive collection of exhibits as well as demonstrations and lectures. It was opened early in 1840 with William Sturgeon as lecturer. Sturgeon may have roused Roberts' interest in electricity. Soon Roberts produced an electro-magnet capable of sustaining a load of 1,400 lbs. He displayed two smaller ones at the Great Exhibition. The Gallery itself failed in 1842.

In about 1832 Roberts had married again. His second wife, Eliza, had three children, Richard in 1833, Eliza Mary in 1835, and John in 1837.[38] At the time of the 1841 census the family lived at Cecil Street, Chorlton-cum-Medlock. Roberts had probably been widowed by 1848, when he was living in lodgings and the younger children had been sent to school in Southport. Richard junior was an apprentice at Roberts' works.[39] Thomas Sharp had died on 21 April 1841 but it was not until 24 June 1843 that the firm of Sharp, Roberts was dissolved by the mutual consent of the remaining partners.[40] No reasons were given why such a successful partnership should have broken up. One suggestion is that Roberts did not wish to specialise in locomotive manufacturing but to use his inventive genius in a wider field. Sharp Brothers retained the Atlas Works and Roberts the Globe Works, described in 1842 as suitable for 'general machine-making, spinning and weaving machinery, and for tools of various kinds'.[41]

Roberts ran the Globe Works alone for twenty-one months, but became ill as well as being lamed through a fall from a horse. In February 1845, he formed a partnership with Benjamin Fothergill as manager superintending the works and Robert Graham Dobinson taking charge of the accounts. Fothergill was forced to retire through ill-health in 1849 or 1850 but the firm carried on as Roberts, Dobinson & Co. until wound up in June 1852. The contents of the Globe Works had been auctioned at a nine-day sale in the previous February. After this, Roberts retained part of the buildings where he had an office as a consulting engineer, moving in about 1858 to Brown Street. During the summer of 1860, he left Manchester for London.

We learn most about Roberts' activities during his final Manchester years from eighteen patents, covering an amazing range of inventions. The first after his separation from Sharp, Roberts was for preparatory machinery for spinning, suggesting that he was trying to develop his own range of textile machinery.[42] Although of little general significance, the patent contained one innovation well in advance of its time, for an automatic stop motion in case of yarn breakage on spinning and doubling machines. Roberts apparently retained rights to the extended patent for his self-acting spinning mule, which had a few years left to run. A patent for improvements to the mule was submitted in 1847, and another in 1854, but again they were of little significance.[43]

Three patents in 1854 show that Roberts spent considerable time and money trying to develop his own design of a combing machine.[44] One such machine was auctioned at the Globe Works in 1852, while another was made for Roberts by William Heywood, who also made for him a cigar-rolling machine. Perhaps Roberts hoped to repeat the success of the mule, but his combing machine failed to compete with those of Donisthorpe, Lister and Heilmann, and there are no accounts that it was ever used in production.

An attempt to mechanise velvet-weaving had similar results.[45] Roberts' patent of 1850 shows his awareness that weaving velvet as a double cloth was quicker, for among the twenty-four inventions was a way of sharpening the cutting knives to separate the two layers of fabric. This patent specified improvements to virtually every part of a power loom, including one for weaving patterned velvet. Roberts' interest in textiles spread to cloth-finishing, for in 1847 he patented improvements to a beetling machine for putting a finish on cloth,[46] and in 1858 patented a complex 'pentagraphic' machine for making or copying multiple images on rollers or flat plates for calico-printing and copper-plate engraving.[47] It is doubtful whether any of these ideas were applied commercially.

Roberts worked for many years to perfect the design of the pendulums on his clocks and carried out experiments to determine the expansion of metals. At the Great Exhibition, he displayed examples of horology from another long patent of 1848 with eight pages of drawings.[48] It covered escapements for watches; a form of stop watch; the use of tides as a winding mechanism to raise weights; a single weight to drive both the striking and going mechanisms; a master clock operating slaves by sending a puff of air down a pipe which Roberts claimed would be a better method of sending messages than the electric telegraph; a form of digital clock in which the numerals were painted on long rolls of cloth and displayed through windows which could be illuminated at night; and much more. This patent confirms that Roberts was equally at home with small delicate mechanisms as with large. Turret clocks continued to be made by Roberts' foreman, John Bailey, and his son W.H. Bailey, after 1852.

The patent of 1847 which included the Jacquard punching machine proved to be a turning point in Roberts' life.[49] The conventional story is that, during the construction of the first tube for the railway bridge over the river at Conwy, the contractors were greatly hampered by combinations among the workmen so that it was feared that the iron work would not be finished in time. 'In their emergency, they appealed to Mr. Roberts, and endeavoured to persuade him to take the matter up. He at length consented to do so, and evolved the machine in question during his evening's leisure – for the most part while quietly sipping his tea.'[50] Unfortunately the veracity of this story must be doubted because the patent application is dated 5 March and construction of the ironwork did not start until some time in June.

The patent drawings show a very complex machine with a single row of ten punches which could be brought in or out of action by a system of Jacquard cards. It appears to have been very difficult to alter the spaces between each punch, which may have limited a more general application of this machine. On the other hand, it was possible to vary the distance which the plate was advanced under the punches. A plate 12ft long could be punched more accurately in four minutes by one mechanic, three labourers and a boy to oil the punches, than the earlier system with a dozen men who would have taken longer. Only two of these machines were ever made. Perhaps Roberts had hoped to use the first for plates of locomotive tenders or boiler plates, but, if for the latter, the holes would have been distorted when the plates were rolled since the Jacquard machine could accept only flat ones. Here we may note that Roberts invented the plate-rolling machine which could bend plates to any curvature simply by altering the position of the middle of three rollers.

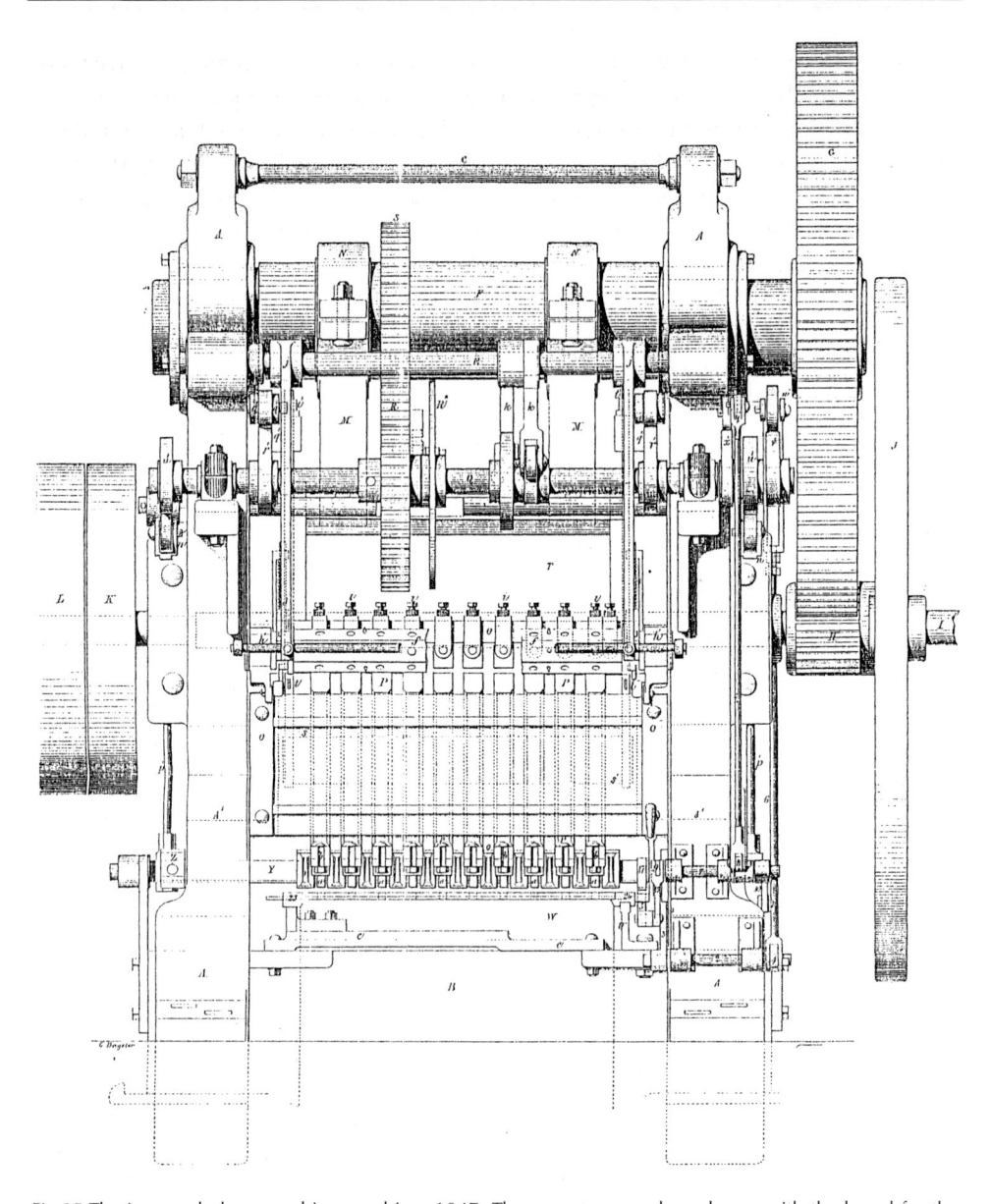

Fig.35 The Jacquard plate punching machine, 1847. There are ten punches, shown with the barrel for the Jacquard cards partly cut away. The plate to be perforated passed through near the bottom. *(Patent 11607, 1847)*

Through his involvement with the Conwy Tubular Bridge, and recognising the strength of beams made on this principle, Roberts conceived the idea of considering a ship as a beam or girder. It was a period when iron was being introduced into shipbuilding but with little or no theoretical knowledge. Roberts took up these ideas enthusiastically and through them lost whatever fortune he still had after the closure of the Globe Works. The first evidence that has survived about Roberts' interest in shipbuilding is found in another vast patent in October 1851 containing nineteen new ideas.[51] Among them were meters for measuring the flow of liquids, water turbines, hydraulic systems for operating lock gates or swing bridges as well as bilge pumps and twin propellers for ships. A turbine and a propeller were displayed at the 1862 International Exhibition. But it was in 1852 that Roberts submitted an even longer patent with thirty-three claims for novelty, mostly concerning the design of ships.[52] Among these inventions were improved interior lighting, better system of ventilation, safer life boats, storage of fresh water and armour for naval ships. The most important ideas were manoeuvrability with twin screws and, above all, cellular construction with longitudinal hollow beams to give exceptional strength. He designed a passenger liner 424ft long, and 60ft broad, to carry 500 passengers. Had it been built, it would have been the largest ship afloat, bigger that Brunel's *Great Britain*, and with cylindrical boilers working at much higher pressures and so more economical with smaller surface condensing engines.

Maudslay's influence is again evident in Roberts' patent of a range of riveting and sheet metal-working machines which could be used in shipyards.[53] The riveting machine could be suspended from an overhead travelling crane so that it could rivet plates to form the sides of ships. Other drilling and punching machines used a template placed below one head with a pointer so that the sheet of metal under a second linked head might be drilled or punched accurately. Many copies could be produced from the one template. The same principle was developed for making watch plates. The punching machines could be adapted for embossing metal plates.

Roberts took models of his ship to the British Association meeting at Hull in 1853, to the Institution of Civil Engineers in 1854 and the Science Museum in London. He went to France in 1855 where he had a personal interview with the Emperor who was impressed with Roberts' inventions but all to no avail. After the dissolution of the Sharp, Roberts partnership, Roberts seems to have worked on his own without any financial backers. He had to bear 'the ridicule and sneers of many at the bare idea of a "mechanical engineer", (a landsman, forsooth!) daring to suggest improvements in the construction of a ship'.[54]

One reason for Roberts moving to London in 1860 was an address Fairbairn gave to the Institution of Naval Architects in which he emphasised the need for stiffening long ships on the principle of a beam without acknowledging Roberts' earlier proposals. Roberts must have thought that he could promote his ideas better in London and became an associate member of this institution. He set himself up as a consulting engineer at 10 Adam Street, Adelphi, round the corner from the institution. Commander Thomas Edward Symonds, a naval officer, helped to promote his ideas, both with the Admiralty and elsewhere. Although in 1862 Roberts' twin screws were incorporated in the *Flora*, a vessel of 400 tons, and some of his other ideas were beginning to gain acceptance, there could have been little financial gain for him.

Roberts' financial position continued to deteriorate. He was looked after by his daughter, Eliza Mary. It may have been Beyer who first realised just how financially distressed Roberts had become. He and other prominent engineers began to organise a subscription while Fairbairn petitioned the Prime Minister, Palmerston, for a civil list pension, but in vain. From the initial money raised a gift of £50 was authorised, but as the result of a fall down a steep flight of stairs, Roberts died in his daughter's arms on 11 March 1864, aged seventy-five. The obituary in *The Engineer* described Roberts as 'one of that class of gifted men whose loss is a national one... Nearly every year of his manhood was signalised by some useful discovery, and even within a few hours of his death his mind was pursuing its wonted track of invention'.[55] His achievement was belatedly recognised by the granting of a civil list pension to his impoverished daughter in 1866.[56]

Endnotes

[1] Bailey 1878-79, p.195. Other biographies may be found in Hills 2002; Smiles 1863, pp.264-72; Dickinson 1945-47; Catterall 1975; Chaloner 1968-69.
[2] *PICE*, 24 (1864-65), p.536.
[3] P.P. 1851 (486) XVIII, Q. 1281.
[4] Smiles 1863, p.265.
[5] See Cantrell 2002.
[6] *PICE*, 24 (1864-65), p.536.
[7] Smiles 1863, p.266.
[8] Pigot and Dean, *New Manchester and Salford Directory* (1817-18), p.205.
[9] Catterall 1975, p.1.5.
[10] *Manchester Guardian*, 5 May 1821.
[11] Smith 1932-33, p.26.
[12] *The Engineer*, 7 (1859), p.207.
[13] Earlier authors claimed that Roberts had twenty-nine patents, but a check of these has revealed that four belonged to another Richard Roberts, in south Manchester.
[14] Patent 4726, 1822; *Manchester Guardian*, 16 November 1822.
[15] Ellison 1886, p.36.
[16] Henderson 1966, p.62.
[17] Hopkins 1824-30.
[18] An account was published in the *Manchester Guardian*, 12 February 1825.
[19] P.P. 1851 (486) XVIII, Q. 1334.
[20] *The Engineer*, 17, 1 April 1864, p.197, confirms his concern for skilled men.
[21] Patent 5138, 1825.
[22] On the development of spinning machinery, see Catling 1970; English 1969; Hills 1970.
[23] Patent 5949, 1830.
[24] Woodcroft 1863, p.39.
[25] Dickinson 1945-47, p.130.
[26] *London Gazette*, 18256 (1826), p.1411.
[27] *Manchester Guardian*, 4 February 1837.
[28] Patent 6258, 1832.

[29] Patent 792, 1856.
[30] T. Sharp to C. Saunders, 25 April 1840: PRO, British Transport Historical Records, CH & I/12/5-6.
[31] These figures are based on the Sharp, Roberts Order Books at the National Railway Museum, York.
[32] Colburn 1871, p.41.
[33] *The Engineer*, 11 (1861), p.188, letter from Roberts.
[34] Fall River Historical Society, William C. Davol's Journal of a Trip to England, 11 February 1839.
[35] Patents 11607, 1847; 990, 1860. The 1847 patent also covered Roberts' Jacquard punching machine, discussed below.
[36] Patents 6536, 1834; 6690, 1834.
[37] Dickinson 1945-47, p.134.
[38] See 1841 census return, Manchester district 35, folio 31.
[39] *Mechanical World*, 23 (1898), p.195.
[40] *London Gazette*, 20237 (1843), p.2163.
[41] Love 1842, p.63.
[42] Patent 10150, 1844.
[43] Patents 11747, 1847; 2272, 1854.
[44] Patents 1591, 1854; 2272, 1854; 2491, 1854.
[45] Patent 12948, 1850.
[46] Patent 11608, 1847.
[47] Patent 488, 1858.
[48] Patent 12207, 1848.
[49] Patent 11607, 1847.
[50] Smiles 1863, pp.271-72.
[51] Patent 13779, 1851.
[52] Patent 14130, 1852.
[53] Patent 1621, 1854.
[54] *The Engineer*, 9, 20 April 1860, p.254.
[55] *The Engineer*, 17, 18 March 1864, p.176.
[56] Dickinson 1945-47, p.134.

5
David Napier
Michael Moss

David Napier (1788-1873) was one of the foremost precision engineers of his day.[1] Improvements which he made to the printing press were adopted throughout the world by the end of the nineteenth century and were not finally abandoned until the introduction of photo-composition in the mid 1970s. Unlike those prominent engineers of his time such as Richard Roberts who were new to the profession and had no antecedents in the metal-working trades, Napier came from a family with deep roots in the handicraft iron-working industries.[2] His contact with Maudslay was almost certainly intended to inform the wider family network of enterprises in London and Scotland, which continued to interact with one another throughout his lifetime.

There can have been few families who made such a significant contribution to so many branches of the engineering industry over such a long period. David Napier's grandfather, Robert (1726-1800), originally managed a family smithy in Dumbarton but this business was transformed into an industrial enterprise by the opportunities presented by the development of the textile trades along the shores of the Clyde during the second half of the eighteenth century. Robert Napier was deeply involved in assisting the mechanisation of calico-printing processes, perhaps through the supply of improved gearing to facilitate the operation of the copperplate presses. Robert extended the engineering connections of his family by marrying into the Denny family, founders of a successful Dumbarton shipbuilding business.[3]

In the late 1760s he was joined in business by his eldest son John (1752-1813) and later by his younger sons David (1756-1828), Robert (1760-1847) and James (1764-1848).[4] John was an enterprising engineer and by 1786 had taken control of the business from his father.[5] By 1790 the family workshop at 93 High Street was equipped with two steam engines: a blowing engine for their cupola, and a rotative Newcomen engine for driving a rudimentary boring machine. The completion of the Forth & Clyde Canal in 1790 gave John Napier direct access to the celebrated Carron ironworks, near Falkirk.[6] No sooner had the canal opened than he was placing more and more business with Carron.[7] He relied on them for supplies of pig iron, foundry tools and for specialist products such as pumps and rollers, although there were often long delays in delivery.[8]

In May 1783 John Napier's brother Robert married Margaret Donald, the daughter of James Donald of Geilston House in Cardross, a Virginia merchant. The marriage cemented the relationship with James Dunlop, of a prominent Glasgow merchant family. He was owner of the Clyde ironworks site, three miles east of Glasgow on the banks of the Clyde, and an alternative source of supply of pig iron. Margaret Donald's brother was married to Dunlop's sister. These connections gave the Napiers access to Glasgow's manufacturing and – significantly for the future of the family enterprises – mercantile communities. Robert and his wife left Dumbarton shortly after their marriage and do not seem to have remained in the west of Scotland.[9] It is possible they went to London to represent the family's interests there and

Fig.36 David Napier (1785-1873) in his old age. *(Napier Power Heritage Trust)*

perhaps for training at the Woolwich Arsenal. While away, they had at least one son, Robert. They returned to Scotland in about 1787 when Robert Napier settled in Inveraray in Argyllshire to become 'smith and armourer' to the 5th Duke of Argyll in partnership with William Simpson.[10] David Napier, the subject of this study, was born there in May 1788. Over the next twenty years Robert Napier and his wife had another ten children, two sons, William (1796) and John (1798), and eight daughters, one of whom, Margaret, married into the Simpson family. At the time of their arrival in Inveraray the Duke of Argyll was engaged in both improving his estate and rebuilding the town. One attraction for Robert was the plentiful supply of iron available from the Argyle Furnace Co. whose works at Furnace on the shores of Loch Fyne had been established in 1754 by the Duddon Company of Lancashire to smelt iron using charcoal.[11] Although this method of ironmaking was fast becoming obsolete, the Argyle Furnace Co. continued in business until 1813. For the family business in Dumbarton it no doubt provided a convenient alternative to the unreliable supply from Carron. The Duke had also set up a woollen mill in 1776 on the Douglas water and Robert Napier's knowledge of textile machinery would have been useful in maintaining and repairing the equipment.[12] Robert Napier was admitted a burgess of Inveraray in September 1806.[13]

Despite being resident in Argyllshire, Robert remained a partner in the family business, suggesting that one of the reasons he was there was to secure supplies of iron not just from Furnace but from the other local ironworks, such as that at Bonawe. David remained at Inveraray until he was apprenticed to the family business at Dumbarton, probably at the age of sixteen in about 1805.[14] With Robert junior and William Simpson already working in the business at Inveraray, and the family growing, there was no room for him at home. Nevertheless the Inveraray business appears to have been substantial. When the family home, the Riding House, was largely destroyed by fire in November 1818, Robert Napier described it 'as admired by all who saw it'.[15]

Late in 1799 John Napier had extended the Dumbarton business to Glasgow to be nearer the Clyde Ironworks which was fully occupied in making cannon. He immediately ceased to place business with Carron.[16] He opened a foundry in Jamaica Street and went to live there himself.[17] His brothers James and David (1756-1828) remained in charge of the Dumbarton works. After their father's death in 1800 the partnership was reconstructed as John Napier & Bros., with the brothers John, David, Robert and James as partners.[18] In that year John became a member of the Glasgow Incorporation of Hammermen and two years later opened a new works in Howard Street, conveniently situated not far from Clyde. Further premises were acquired in Tradeston during 1806 in a complicated deal involving some of the best known engineers and manufacturers in Glasgow including the engineer James Robertson; James Sword junior, ironmonger; William Watt, wright; John Smith, machine-maker; William Dunn, cotton machine-maker and spinner; and James Cook, a well-known engineer in Tradeston who owned the site.[19] During that year both James and Robert became Glasgow burgesses. With the resumption of the war with revolutionary France in 1804, both parts of the business must have been largely devoted to the boring of cannon during David Napier's (1788-1873) apprenticeship. This would have lasted for five years until 1809, when he qualified as a blacksmith and became a journeyman. If his path followed that of his cousin Robert (1791-1876), son of James, he would have gained considerable knowledge of calico-printing machinery.[20] With only three years between them, the two cousins became close friends. Although Robert formally started his apprenticeship sometime after David, he was already working occasionally in the business. His cousin David (1790-1869), the son of John, was also in the Glasgow shop – 'I never served a regular apprenticeship to anything but put my hand to everything, and by the time I was twenty years of age I had complete charge and control of my father's business in every respect'.[21]

David Napier's (1788-1873) career immediately after completing his apprenticeship remains unclear. Wilson and Reader, in a book commissioned to mark the 150th anniversary of the enterprise, claimed he started out on his own in London in 1808.[22] Letters which have subsequently come to light show that this is incorrect, although it still remains uncertain when he decided to go south. In any event David Napier seems to have been in London by 1814 working for Maudslay with the intention of returning with his new found knowledge to work in the family enterprises in Glasgow. The fact that Maudslay does not seem to have objected to the Napiers learning from his skill suggests that they were already known to him. A letter written by David Napier to his cousin Robert in April 1817 gives the impression that much of his time with Maudslay was devoted to building presses to the firm's own designs and making cast steel dies for pressing silver spoons. His cousin was very keen to have details of the press but David Napier, on his own admission a dilatory correspondent, forgot to send them to him. By the time he did, he had left Maudslays and had 'to apply to one of Maudslay's Men first who is something like myself very slow in their performances'. The letter includes a detailed drawing of the press and a description of the method of working. In the same letter he also provided details of Maudslay's 'triangular bar laith [lathe]' although he continues 'I would not recommend them to you as they are very expensive; but I would recommend a cast iron bed laith'. He declared a preference for John Jacob Holtzapffel's headstock over Maudslay's design.[23]

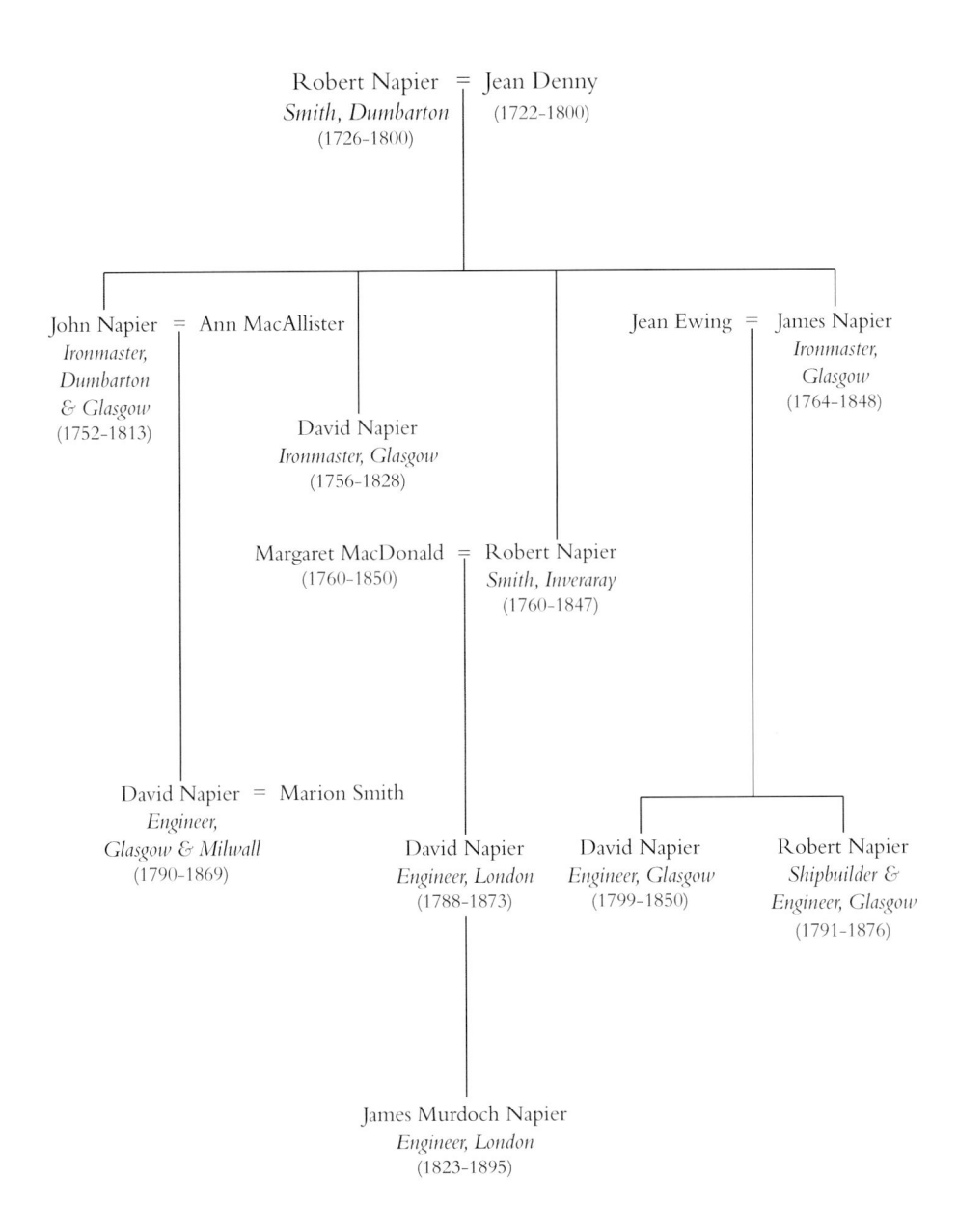

```
                    Robert Napier  =  Jean Denny
                    Smith, Dumbarton   (1722-1800)
                    (1726-1800)
                            |
      ┌─────────────────────┴──────────────────────────────┐
John Napier = Ann MacAllister                    Jean Ewing = James Napier
Ironmaster,                                                   Ironmaster,
Dumbarton                                                     Glasgow
& Glasgow              |                                      (1764-1848)
(1752-1813)       David Napier
                  Ironmaster, Glasgow
                  (1756-1828)

            Margaret MacDonald = Robert Napier
            (1760-1850)          Smith, Inveraray
                                 (1760-1847)
                    |
   ┌────────────────┼───────────────────┬──────────────────┐
David Napier = Marion Smith
Engineer,
Glasgow & Millwall    David Napier      David Napier     Robert Napier
(1790-1869)           Engineer, London  Engineer, Glasgow Shipbuilder &
                      (1788-1873)       (1799-1850)      Engineer, Glasgow
                            |                             (1791-1876)
                      James Murdoch Napier
                      Engineer, London
                      (1823-1895)
```

Fig.37 An abbreviated Napier family tree.

Fig.38 The drawing of the press manufactured by Maudslay, which David Napier sent to his cousin Robert during his early years in London. *(GUA)*

By April 1817 David Napier had for the past eighteen months been foreman in an unidentified factory at 8 Nevil's Court, Fetter Lane, off Fleet Street,[24] from which it can be calculated that he had left Maudslay by the end of 1815. At his new employer, 'the principal of their tread [trade] is printing press machines – but as I have always the best of the work I am seldom on them as there is generally some other fancy machines making & in the doing of which they are always foolish enough to late me take my own way'. Although he said it himself, his skill as a precision engineering craftsman of outstanding ability was already well recognised. He explained to his cousin Robert 'the foolish manner' in which he had been spending his evenings:

> *I betook myself to invention & invented a machine to supersede the necessity of climbing chimney sweeps, but after being at considerable trouble & expense in making of a model I found it was much too complicated for the sooty mechanics into whose hands they would be put & the soot would be very hurtful to them for which reason I layd it to one side. I then invented a machine for drawing volutes* [helixes] *(either plain or oval right & left & to any size) this instrument I also maid but when I came to try it I found that it would require considerable alterations which I could not attempt …*

He reported that he had tried to sell the idea to mathematical instrument makers but had found little interest due to 'tread being so dull'. With these and other failures behind him, he had turned his attention to the printing press. He boasted:

> *I have invented a very grand improvement uppon the printing press which would not at present sell for two hundred pounds for could I only get money to begin with I have no fear of plenty of work for I know that my improvement would highly please the printers.*[25]

He wrote in June to his cousin David Donald stating he wanted 'to procure a good business in London in a Good Tread which is very difficult to be found in these critical times'.[26]

This change of heart was a consequence of his uncle John's death in 1813 just as the Glasgow works was in the process of expanding. John's son David (1790-1869) had won the contract in 1812 to make the boilers and castings for the engine of Henry Bell's steamboat *Comet*. According to his account, 'seeing steam navigation was likely to succeed, I erected new works at Camlachie for the purpose of making steam-engines'.[27] These works in the east of the City were near to supplies of coal and iron and close by the foundry and engine shop of Francis Smith, to whose daughter David Napier was engaged. Trading under the name of John Napier & Son, they were operational by January 1815 when David Napier assisted by Duncan McArthur, already an established marine engineer in Camlachie, recruited Joseph Hill as his boilermaker. Hill was shown over the works by David Napier accompanied by McArthur and saw 'a punching machine of new construction that was then making in the works'. Initially Hill was doubtful if it would work but on inspection was more than convinced. It is tempting to think that its design and conception owed much to information gleaned by David Napier (1788-1873) in Maudslay's works.[28] In September 1816 the old partnership of John Napier & Bros. was dissolved and the Howard Street works abandoned.[29] The Tradeston works, which were managed by John's son Robert, were sold to the well-known Glasgow engineer James Cook.[30]

There was now no family business in Glasgow for either David or Robert to work in. In 1818 Robert seems to have left the Carmyle works and set up on his own account in Greyfriars Wynd.[31] It must nevertheless have come as a surprise when Robert Napier received a letter from his cousin David (1788-1873) early in 1818 from 11 Portman Buildings in Soho Square:

> *I have layd aside all thoughts of returning to remain in Glasgow as I have now begun business with very flattering hopes of success in partnership with Mr Baisler a stationer in Oxford Street who has found out several ingenious inventions. I have now five men at work besides myself & I expect in a very short time to be able to imploy a few more ... untill that we get a few patent machines brought forward what we at present propose is Engine Laiths & Printing Press manufacturing.*

He added he was teaching at the Sabbath school in Swallow Street Church – St Anne's Church of Scotland congregation in Soho.[32] Mr Baisler was Francis Baisler, a partner for about a year with John Murray in a book-selling and printing business at 326 Oxford Street.[33] One of the 'ingenious inventions' David Napier referred to in his letter was an improvement in the method of cutting paper, known as the Baisler patent paper plough which was patented in 1817.[34] An early acquisition of the new firm of Baisler & Napier was a lathe from Maudslay, which was still working over fifty years later.[35] In 1820 when an American Daniel Treadwell patented a treadle printing press, Napier built it for him and at about the same time constructed a working prototype of Rutt's single cylinder printing machine.[36] He moved to larger premises at 15 Lloyd's Court, Crown Street, Soho in 1820-1.[37] The partnership between Baisler and Napier was dissolved in April 1822 and Napier continued in business on his own.[38]

During these early years Napier himself invented a folding pocket drawing compass and tracing instrument, which he called the 'Universal Perspectigraph' to 'afford a more accurate method of copying outlines of any kind such as drawings, maps etc., either right or reversed, on copper for the engraver'. He claimed it was much better than other contemporary pantographs.[39] He submitted it to the Society of Arts in January 1819 and was awarded a premium of ten guineas. This device reflected his knowledge of calico-printing as pantographs were essential for the accurate copying of repeat patterns on to the engraved rollers. David Napier's first venture into manufacturing printing machines to his own designs seems to have been around 1820 when he built a press whereby one man both inked the roller and performed the printing process. This presumably is the invention he referred to in 1817. Despite the claim in his advertising that 'one man at this machine would be more than equal to two at the common press', he sold only two. The Society of Arts, which examined his invention, declared unhelpfully 'its merits ... by no means substantiated the assertions of the circular'.[40] The drawback of the press was the length of time it took to set up for each job, which would not have been a problem in the calico trade where long runs were common and small batch work was carried out by block printers.[41]

This rather public setback did not discourage Napier, who began work on a large perfecting machine to print on both sides of the paper in one operation with a pair of cylinders. This demanded all his ingenuity as the register of the type had to be the same on one side as the other to allow for the binding margins, and inking had to be carefully controlled to give clean printing. He overcame these problems by improving the means of guiding the paper through the press and introducing the rising-and-falling or rocking cylinder to keep the cylinders away from the type until the moment the impression was made. In most printing machines the paper was guided by tapes which ran in the margins. They had to be altered if the margins changed, requiring skill and running the risk of the tapes breaking. Instead of tapes, Napier introduced what Thomas Hansard, the Parliamentary printer, later described as a 'beautiful mechanism ... contained in the interior of the impression cylinders, which have openings along their circumference, through which the *grippers* perform their operation, and upon their action depends that important desideratum of press-work, accurate register, or the backing of pages on the paper, and this purpose is so fully effected, that from the many thousands of sheets which have passed through my machine, without the smallest deviation after register was made, I venture to call them infallible'.[42] Napier's other invention, the rising-and-falling or rocking cylinder, caused the cylinders to rise after each impression allowing the forme of type to slide back untouched and then to fall back in time to take the next blank sheet of paper. Hansard eulogised this device which he believed 'reflects the highest credit upon the mechanical skill of the inventor ... for it is principally owing to this singular contrivance that he has been enabled so wonderfully to compress and simplify his machine as to bring it within

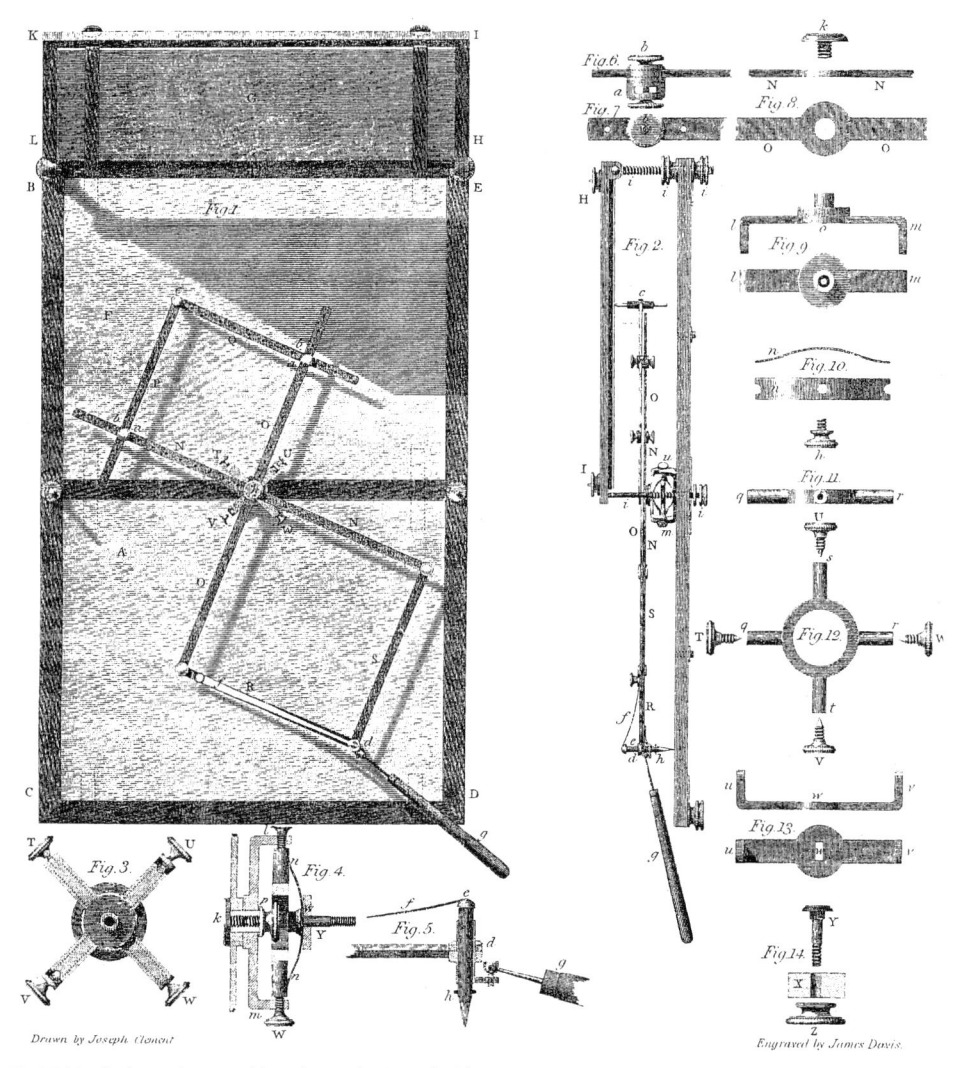

Fig.39 Napier's tracing machine drawn by Joseph Clement. *(TSA)*

the capability of so small a power to produce so much work'.[43] For all its sophistication the press was hand driven by a two handled crank, which Hansard considered contributed to it being 'more likely to succeed in all its pretensions than any which has yet been offered to us; more especially as it supersedes any necessity of steam power'.[44]

Napier christened his invention *Nay-Peer* and no sooner had the design been perfected than he sent a set of drawings to his cousin Robert in Glasgow. In a covering letter he regretted that his cousin David (1790-1869) had failed to find him at home on a visit to London 'on account of his being rather Chieftain of the Clan so that he might honour the name by standing sponsor at the font and drinking a bumper to the *Nay-Peer* printing machine which was at that time upon my premises but now on its way to Paris'.[45] The first press was exported to Paris, a centre of fine printing, and from 1824 Napier made several for Hansard who reckoned that if properly managed they could print 1,000 sheets an hour. Despite its name and Hansard's enthusiastic endorsement, few were sold. There are no surviving sales records of printing machinery until 1832 when a further machine, by this time steam driven, was sold to Hansard.[46] A final *Nay-Peer* was sold the following year to a customer in Derby and thereafter none are recorded. It was replaced by a new model perfecting machine of which no fewer than eighty-seven were sold between 1836 and 1863 at prices ranging from £450 to £600. They were mostly steam driven and largely supplied to general printers, or printers specialising in quality products such as bank notes or fine editions. During the 1850s perfecting machines were supplied to Waterlows in Edinburgh, who printed bank notes; the playing card printers De La Rue; Bradbury Evans; and the Queen's printers Eyre & Spottiswoode. Ingram & Cook, who founded the high quality weekly publication *Illustrated London News* in 1842, purchased a machine in 1844.[47] During these years the newspaper industry had no use for perfecting machines because of composing constraints.

According to Hansard, however, Napier had counted newspaper proprietors among his clients from at least 1822. He supplied two models to the newspaper trade, the *Desideratum* or *Imperial* and the *Double Imperial*, which, although the inking arrangements were almost

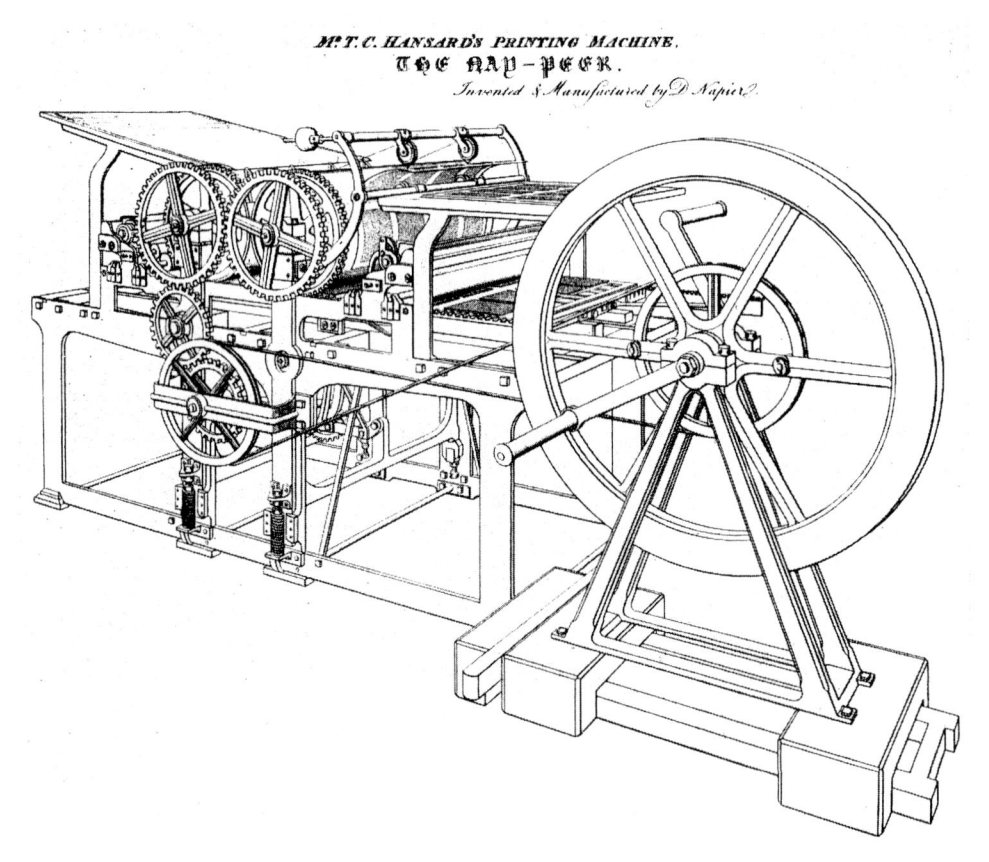

Fig.40 The *Nay-Peer* press drawn by David Napier for T.C. Hansard's book *Typographia*, published in 1825. The crank drive allowed the press to be operated anywhere, but it was hard manual work.

Fig.41 A power-driven perfecting printing machine in operation in about 1828. *(Napier Power Heritage Trust)*

identical to those in the *Nay-Peer*, were designed to be simpler to set up and to achieve high throughput. The *Desideratum* was a single-cylinder machine, printing on one side of the paper only, while the *Double Imperial* also printed on only one side but as the name suggests had two cylinders which allowed the type forme to print as it moved in both directions. The *Double Imperial* reportedly could print 2,000 sheets an hour. The most popular was the *Desideratum* of which at least 140 were sold, whereas the *Double Imperial* achieved sales of only twenty-three or so. By 1825 the *Morning Post* and the *News* were being printed on the *Imperial* and the *Courier* and *British Traveller* on the *Double Imperial*. The *Courier* had installed its press in 1823 and at once announced to readers:

> *We think it right to announce to our readers that the Courier is now printed by a machine of such extraordinary power that it is capable of throwing considerably above two thousand papers per hour; it has, indeed, on one occasion, produced at a rate of 2,880 impressions within the hour.*[48]

Napier also turned out about half a dozen much bigger presses, known as the *Large Quadruple*, a further modification of the *Desideratum*, two of which were supplied to the Liberal newspaper *Morning Chronicle* and one exported to New York in 1838 for the *Courier Enquirer*. His principal customers for his newspaper presses were local papers, which were being established in unprecedented numbers up and down the country.[49]

During these years David Napier was in close touch with his cousin Robert in Glasgow and with Robert's son James R. Napier.[50] David's sister Helen had married Robert's younger brother, also David (1799-1850).[51] He too was an engineer, working with Robert at the Camlachie Foundry, which Robert had leased in 1821 when his shipbuilder cousin David (1790-1869) had moved to the larger and more conveniently situated Lancefield works on the north shore of the Clyde. Robert's ambition at Camlachie was to follow his cousin into the marine engineering trade. After building the engines for the *Leven* in 1823 his business never looked back and orders flowed in. From the outset he appreciated that quality and reliability were the hallmarks of success.[52] In 1826 he built on his own account the PS *Eclipse* and wrote to his cousin David (1788-1873) in London to enquire if he could find a buyer. He described her in glowing terms 'as the most complete vessel of her size ever built on the Clyde; in point of sailing unequalled by any vessel; built of the best British oak, copper-sheathed and fastened, with double side-lever engines, having cylinders 35in in diameter, warranted equal in construction and workmanship to the best engines made'.[53] The following spring, Robert Napier travelled to London to stay with his cousin to explore the market for his engines and if possible to see over the Maudslay works. Introduced by his cousin, Maudslay responded in the most generous terms:

> Mr Maudslay's respectful compliments to Mr Napier, and begs to say he always feels more gratification in meeting or seeing any gentleman who has a knowledge of the business he is engaged in than the thousands who go about taking up the time without gaining any information ... Mr M. will therefore be glad to see Mr N. either on the receipt of this or at 4 o'clock, or tomorrow morning Friday.[54]

It is tempting to speculate that Robert Napier put what he saw to good effect; perhaps when his firm became a serious competitor for naval contracts Maudslay regretted his hospitality. There is certainly some circumstantial evidence for such a point of view. Robert Napier placed the same emphasis as Maudslay on the quality of his machining and the reliability of his engines. When his engines for the *British Queen* (1839) were compared with the Maudslay *Great Western* engines, they were found to be much more robust, even though the latter's were more powerful.[55]

The connections between David Napier (1788-1873) in London and Robert's family continued. In 1831 he collaborated with two of Robert's brothers, James and William, to patent improvements to the steam carriage invented by their cousin David (1790-1869) to carry passengers from Dunoon to Loch Eck. On cousin David's own admission this had failed due to 'the softness and the hilliness of the roads, and more particularly from want of knowledge how to make a boiler ...'.[56] James Napier, the inventor of the tubular boiler and owner of the Swallow foundry in Glasgow, no doubt was responsible for the modifications to the boiler design of their new patent. The other feature was the use of belt drives rather than direct action to transmit the power to the axles.[57] Despite these improvements the carriage was no more successful than its predecessor.

On Friday 25 July 1835 David Napier's (1790-1869) steamer *Earl Grey* was at the quay at Greenock preparing to race Robert Napier's *Clarence*. She had recently been re-boilered[58] and as the steam pressure was raised just before the start, she exploded with 'dreadful loss of life'.

David Napier

'A great many people who were standing on the quay were forced up in the air and fell down dreadfully injured'.[59] Although he was exonerated in the subsequent enquiry, this accident had a profound effect on David Napier (1790-1869) who at the end of the year leased his Lancefield works to his cousin Robert, and moved to a yard on the Thames at Millwall in London. At about the same time David Napier (1788-1873), in need of more space, moved his works from Soho to York Road, Lambeth. The premises were not large, but they were by all accounts well-equipped and employed between 200 and 300 men.[60] A visitor in the 1860s commented that 'the workmen were well-qualified ... as some of the machines they made were as delicate as any clock could be'.[61] A printing machine after all had to be accurate to one sheet of paper. In 1837 David Napier's eldest son James Murdoch, then aged fourteen, joined him as an apprentice in the business. It is not known what if any contact there was

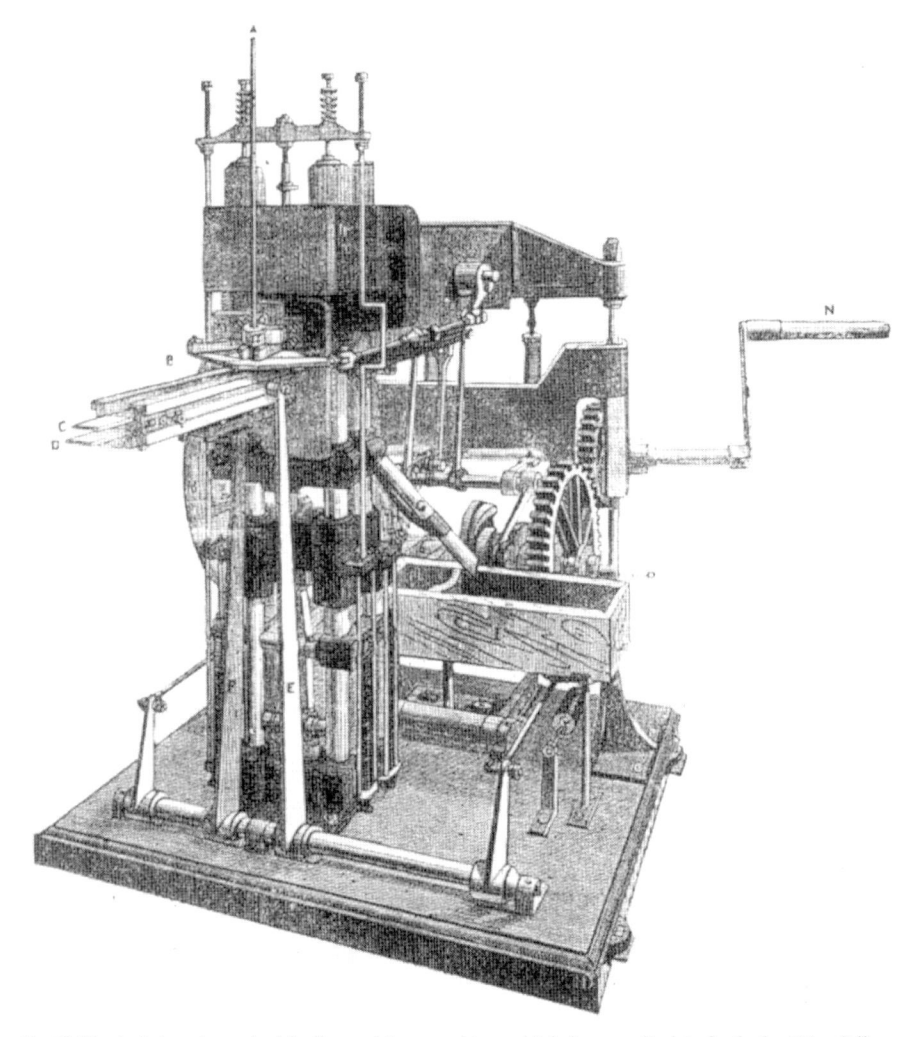

Fig.42 Napier's hand-cranked bullet-making machine which he supplied to both the War Office and foreign governments. *(Napier Power Heritage Trust)*

between the two Davids in London. David Napier (1790-1869) makes no mention of his namesake in his autobiography.[62]

In the late 1830s the Board of Ordnance invited David Napier (1788-1873) to design and build a machine to press bullets out of strips of lead rather than cast them, which had been the practice in the past. Pressed bullets had the advantage that they did not contain air bubbles to disturb their trajectory. The resulting hand-cranked press could produce some 25,000 bullets a day with twelve men and five boys employed for continuous operation. Initially David Napier kept his invention to himself, simply supplying bullets to the Board. This position was untenable after the Board placed an order for the machine itself. Consequently he patented his device and began supplying bullet-making equipment to other governments. So impressed was the Board with the quality of workmanship that orders for other equipment for Woolwich Arsenal followed, including the Board's first steam powered gun-finishing machinery. This took Napier back to his early days working at Dumbarton in the family business and led to a growing export business in arsenal equipment, starting with Spain in 1848.[63] An order from

Fig.43 The hydraulically driven *Automaton* coin-weighing and classifying machines. *(Napier Power Heritage Trust)*

Russia the following year for the arsenal at St Petersburg included six gun-boring and turning machines and two trunnioning machines along with other tools.[64] Ironically within five years this plant was being used to manufacture ordnance to fight the British and their French allies in the Crimea. It was perhaps a sense of remorse which persuaded Napier during the war to bore cannons at his own works to relieve pressure on the arsenal.

The connection with the Board of Ordnance led indirectly to further diversification when in 1842 Peter Ewart, chief engineer and inspector of machinery for the Royal Navy, recommended David Napier to his friend William Cotton, deputy governor of the Bank of England. Cotton had been working for some time on developing an accurate weighing machine which could detect if a gold coin had become so worn that it was no longer legal tender. With assistance from Napier, Cotton patented his device. The first prototype weighing machine was built for the Bank of England in 1842 and orders for three more quickly followed.[65] The great advantage of the balance was that it moved upwards only if coins were of incorrect weight; otherwise it remained in equilibrium and the coins were dislodged by a metal tongue (designed by Napier) which moved from a recess. This mechanical weighing machine, known as the *Automaton*, could sort 10,000 coins in six hours while only 6,000 could be processed on hand scales. During the 1840s David Napier and his son also added hydraulic engineering to their range of specialities, supplying hydraulic traversing frames for moving wagons from one line to another at Great Western Railway's Bristol Temple Meads station and machines for forcing railway wheels on to axles.[66]

They were also consulted by William, 3rd Earl of Rosse, an astronomer and himself an innovative engineer, for help with the precision engineering of his great reflective telescope which was erected at Birr Castle in King's County (now County Offlay) in Ireland. David recommended subcontracting the cutting of the screws for raising and lowering the great tube, which was large enough for a man to walk through with a raised umbrella, to Joseph Clement.[67] He had been Henry Maudslay's chief draughtsman, but was reputedly the most skilled gear cutter of his generation. After leaving Maudslay he had set up in business on his own account in Newington Butts. The screw for the Earl's telescope was a foot wide and given the delicacy of the instrument had to be cut with great accuracy. When the Earl received the bill for £400, he protested and threatened legal action. Napier proposed arbitration and when one of Maudslay's sons said he would happily give £400 for such an excellent piece of workmanship no more was heard about the matter.[68]

In 1847 David Napier took into partnership his second son James Murdoch Napier, who had been working in the business for a decade, renaming the firm David Napier & Son.[69] James Murdoch was by all accounts an even more talented engineer than his father and a consequence of his becoming a partner was that patents were taken out in their joint names. David Napier still acted as an agent for Robert Napier in London and also carried out ship repairs for him there. In 1857 he asked for an estimate for three steamers and a number of towing barges, apparently to bring fresh water in to London.[70] This in turn led to collaboration between David and James Murdoch and Robert's eldest son, the brilliant engineer James R. Napier, in the design of what was termed the 'captain's registering compass'. This ingenious device, which was patented in the joint names of David and James Murdoch in 1848, recorded the ship's course on paper so that the captain would have an accurate record for his log.[71] It was supplied to Robert Napier himself and to at least two of his customers,

Fig.44 James Murdoch Napier (1823-1895). *(Napier Power Heritage Trust)*

P&O and the Royal Mail Steam Packet Co. When Robert Napier was negotiating the sale of marine engines to the Russian government in 1849, a compass was presented to the Tsar's brother, the Lord High Admiral of Russia. Although David Napier was proud of the registering compass and despite the fact it won a prize medal at the Great Exhibition in 1851, it did not sell in large numbers.[72] It was later tested by J.R. Napier on HMS *Hyacinth* in the 1860s.[73] Regular contact with Robert Napier and his son continued well into that decade.[74] David Napier also kept in touch with his Argyllshire family. When his parents died within a month of each other in 1845, he and his younger brother William erected a granite monument to their memory in the Old Kilmalieu burial ground.[75]

Increasingly James Murdoch Napier resolved technical problems and made modifications and improvements to the firm's products. When in 1851 Sir John Herschell, the Master of the Mint, placed an order for a re-modelled Cotton weighing machine to test new coins, it was he who undertook the task.[76] By this time the staple market for newspaper printing machinery was beginning to be threatened and eventually eclipsed by the development of rotary presses. Although David Napier had patented a sheet fed rotary press in 1837,[77] he had been unable to persuade newspaper proprietors, who blamed their existing presses for late delivery, to invest in them. Moreover by then *Imperial* and *Nay-peer* presses, which had never been patented, were being widely copied. Against this background he and later his son James Murdoch Napier concentrated on developing platen (flat bed) presses for their fine printer customers. After experimenting with a double cylinder machine, the Bank of England in 1853 placed an order for two double-ended platen machines designed by J.M. Napier to print bank notes. With a unique inking device, these presses could print 1,500 notes an hour at an eighth of the cost of the hand presses which they replaced. The patent for 'inking apparatus for platen printing machines' was taken out in 1853 in the sole name of James Murdoch Napier and thereafter with one exception all the firm's patents were in his name.[78] Apart from their established domestic customers, Napiers won export orders for their platen press from Russia, Germany, Italy and even Japan.[79] Further modifications were patented in 1857 and 1859.[80] James Murdoch also modified his father's bullet-making machinery[81] and, presumably on the invitation of customers, devised solutions to a variety of technical problems including improvements to furnaces used in the manufacture of soda, apparatus for paying out submarine telegraph cables, and machinery for the manufacture of sugar.[82]

During the late 1850s his father, now in his early seventies, moved out of his house next to the works in York Road and went to live in Victoria Road, Surbiton, with his wife, daughter Isabella and son John, a civil engineer.[83] He called his house Dunedin Villa after property he had purchased in Dunedin in New Zealand for his sons David and Peter.[84] He went to the works each day to read the newspapers, and after his wife died, returned to live with his son and family in Lambeth.[85] He must at times have had to resume the reins such as in 1861 when his son was absent in St Petersburg for nine months installing equipment in the Russian Mint.[86]

David Napier, like other members of his family, was a talented engineer with no head for figures.[87] He was more interested in the technical challenge of a contract rather than if it would make a profit. This was reflected in the works, which was equipped with the most advanced machine tools irrespective of cost. Bookkeeping was not one of his strong points, and reputedly between 1853 and 1860 his works accountant failed to make any entries in the cash book, preferring instead to keep a diary. He was dismissed and a young bookkeeper, William Parminter Cardell, was recruited from an ironworks at Spennymoor.[88] Towards the end of Cardell's life, he recorded his career in an unusual autobiography.[89] His first impression

Fig.45 An ornately mounted Captain's Registering Compass. *(Napier Power Heritage Trust)*

was that David Napier and his son James, whom he referred to as Pater and Filius 'were remarkable men, being much above the average of clever persons'. He considered, however, that if 'Pater was a clever man, Filius was undoubtedly a cleverer one'.[90] He described David as by now 'of spare build about eighty years old, of good height, upright, a Scotsman by birth. His manner was always grave and non-communicative'. It did not take him long to find out that 'the firm did not profess to sell at a low price, but they did maintain, and with very good reason, that they always used their best efforts to give customers an equivalent for their money'.[91] With a free hand to reorganise the firm's bookkeeping systems, it soon became clear to Cardell that, despite such high prices, jobs were not always profitable.[92] With an understanding of cost accounting, he put in place a system for 'determining the gross and net outlay on each order or piece of work undertaken no matter of what magnitude or for what purpose, month by month, and exposing it to view in a ledger ruled to accommodate such exhibition'.[93] He observed of J.M. Napier: 'He put honesty, truth and perhaps pride, in front of all pecuniary questions. Such a man could not expect to make a colossal fortune'. He added that his honesty did not extend to dealings with the Inland Revenue: 'His objection to government officials knowing a man's profit was insurmountable. Consequently, he did not blame a person making a spurious return of his income'.[94] After Cardell had been with the firm for six years, David Napier formally retired in December 1866, leaving James Murdoch Napier in total control of the enterprise. Nevertheless he does seem to have assisted in the design of some of the firm's inventions – not perhaps to a method of 'serving mustard through an orifice in lieu of the present inconvenient mode of serving it by means of a spoon' or 'wine glasses to keep insects out of wine', but almost certainly to improvements in intaglio printing machines, his weighing machines and to the 'doctors' used in calico printing equipment.[95] David Napier died in Lambeth on 17 June 1873 aged eighty-five.

Endnotes

[1] His birth is usually given as 1785 but this is now known to be wrong as his christening is recorded in the Inverary parish registers in 1788, the surname misspelt: see http://scots.origins.net. The main published source on Napier is Wilson and Reader 1958. Since their book was written, Maj. Andrew Napier, a direct descendant of Robert Napier (1791-1869), has collected many family papers and deposited them with Glasgow University Archives (GUA) where they form DC 90. Certain papers available to Wilson and Reader appear to have been lost when the business was acquired by English Electric in 1942, which itself subsequently merged with GEC (later Marconi plc and now forming part of Alstom Napier Turbo-Chargers Ltd of Lincoln). Some records of David Napier & Sons are now in the custody of the Napier Power Heritage Trust in Hammersmith. Maj. Napier and Graham Kennison compiled a comprehensive family tree, available at GUA, which forms the basis of much of the family information in this chapter. See Kennison and Napier 2000; also Cardell 1912.

[2] Both David Napier (1790-1869) and his cousin Robert (1791-1876), the shipbuilder, were immortalised in hagiographies which for all their shortcomings contain much useful information about the Napier family: Napier and Bell (eds.) 1912; Napier 1904.

[3] Lyon 1975, introduction; *Statistical Account of Scotland*, 4, p.23, and 3, p.446; Macleod 1884, pp.134, 144. Carron & Co. records confirm Robert Napier's deep involvement in assisting with these developments: NAS, GD 58, Carron collection, 6/1/23-35, letter books 1787-99.

4. Napier 1904, p.2, states that only Robert and James entered the firm but the partnership also seems to have included David (1756-1828).
5. Deduced from NAS, GD 58, 4/15/2, accounts current ledger 1786-90, p.518.
6. See Campbell 1961.
7. NAS, GD 58/4/15/2, accounts current ledger, 2, 1786-90, p.518, records a very low level of business whereas 4/15/4, accounts current ledger, 2, 1790-95, records a more substantial trade.
8. NAS, GD 58/6/1/26-31, indexed letter books 1790-96.
9. A search of the old parish registers for Dumbarton and its surrounding parishes, the Glasgow parishes along with those on the south shore of the Clyde, reveals no trace of them.
10. Headstone of J.R. and W. Simpson, Old Kilmalieu Burial Ground.
11. Hamilton 1963, p.192, gives the date of foundation as 1775 whereas the *Statistical Account of Scotland*, 5, p.298, gives 1754.
12. *Statistical Account of Scotland*, 5, p.297.
13. *The Burgesses of Inveraray 1656-1963*, p.201.
14. Although his cousin Robert Napier's apprenticeship indenture was drawn up after his eighteenth birthday it was backdated to his sixteenth birthday: Napier 1904, pp.5-8. His cousin David (1790-1869), who never undertook a formal apprenticeship, began working for his father at the age of twelve: Napier and Bell (eds.) 1912, p.16.
15. R. Napier to D. Napier (1790-1869), 2 December 1818: GUA, DC90/2/4/5.
16. Deduced from NAS, GD 58/1/21/36-42, letter books, 1799-1813.
17. *Glasgow Courier*, 11 December 1799.
18. *Edinburgh Gazette*, 17 September 1816, p.294.
19. *Register of sasines for Glasgow*, No.6204, 1806.
20. Napier 1904, p.10.
21. Napier and Bell (eds) 1912, p.16.
22. Wilson and Reader 1958, p.13.
23. D. Napier to R. Napier, 14 April 1817: GUA, DC 90/2/4/3.
24. A search of London directories has not revealed the identity of this firm.
25. D. Napier to R. Napier, 14 April 1817: GUA, DC 90/2/4/3.
26. D. Napier to D. Donald, 16 June 1817: GUA, DC 90/2/4/2.
27. Napier and Bell (eds) 1912, p.17.
28. NAS, CS 239 N8/5, advocation David Napier v Hill, 1816.
29. *Edinburgh Gazette*, 17 September 1816, p.294.
30. *Register of sasines Glasgow*, Nos 9698, 1814 and 9933-34, 1815.
31. *Glasgow Post Office directory*.
32. D. Napier to R. Napier, 5 January 1818: GUA, DC90/2/4/4.
33. Brown 1982, p.134.
34. Patent 4180, 1817; the author is grateful to Geoff McGarry for locating this patent.
35. Cardell 1912, p.327.
36. Hansard 1825, pp.698-99.
37. Todd 1972, p.7.
38. *London Gazette*, 8 April 1822.
39. Wilson and Reader 1958, pp.17-18. It can be seen on display in the Musée de l'Impression sur Étoffes at Mulhouse in France.

[40] Hansard 1825, pp.657-58.
[41] Turnbull 1951, p.103.
[42] Hansard 1825, pp.712-13.
[43] Hansard 1825, p.712.
[44] Hansard 1825, p.713.
[45] D. Napier to R. Napier, 26 December 1822: GUA, DC90/2/4/7. It is conceivable that this machine was not the later Nay-peer as Colin Whitehead's research suggests that this may have been a newspaper printing machine.
[46] The sales ledger for printing equipment from 1832-1844 is now held by the Napier Power Heritage Trust.
[47] Wilson and Reader 1958, p.22.
[48] Hansard 1825, pp.713-14.
[49] Wilson and Reader 1958, pp.22-24.
[50] Although David Napier does not seem to have had regular contact with his cousin David the ship-builder (1790-1869) after his move to Millwall, the shipbuilder's sons were certainly well-known to Robert. The eldest John D. Napier was assistant manager in Robert Napier's shipyard from 1855 and, after Robert fell out with James R. Napier, principal manager from 1857 to 1863: GUA, Napier and Kennison family tree, p.3.
[51] David Napier later purchased in 1837 the Parkhead Forge in the east end of Glasgow from the Reoch Brothers to supply his brother's forge and shipyard. He became bankrupt in 1847 and died in 1850: see Hume and Moss 1979, pp.14-17, where he is confused with his cousin David Napier, the ship-builder (1790-1869).
[52] Napier 1904, pp.29-33.
[53] R. Napier to D. Napier, 3 November 1826: GUA, DC90/2/4/8.
[54] Quoted in Napier 1904, pp.34-35. This letter, dated 1 March 1827, is not among the Napier family papers.
[55] Napier 1904, p.117; Rolt 1957, p.99.
[56] Napier and Bell (eds) 1912, p.21.
[57] Patent 6090, 1831.
[58] Napier and Bell (eds) 1912, p.58.
[59] *The Times*, 28 July 1835, p.5.
[60] Wilson and Reader 1958, pp.25-27.
[61] Wilson and Reader 1958, p.27.
[62] Anon. 1904. David Napier effectively retired from the Millwall business in the early 1840s and devoted his talent to inventions and managing Channel steamers. He died in 1869.
[63] Wilson and Reader 1958, pp.29-30.
[64] Wilson and Reader 1958, p.48.
[65] Wilson and Reader 1958, pp.32-33.
[66] Wilson and Reader 1958, p.34.
[67] See Chapter 6.
[68] Cardell 1912, p.305. Clement faced a number of arbitrations relating to his supposedly exorbitant charges: see pp.105-106.
[69] Apart from John, the other children chose different careers: the eldest, Robert, in America where he fought in the Civil War; William Donald became a dentist, and David John and Peter Murdoch

70 farmers, although Peter had originally trained as an engineer: Napier and Kennison 2000, p.6.
71 GUA, DC 90/2/4/42, p.9.
72 Patent 12220, 1848. The National Library of Scotland holds a bound volume of James Murdoch Napier's patents, 1848-1872 (shelfmark L.201.d) which was presumably the property of his father.
73 Wilson and Reader 1958, pp.35-37.
74 GUA, DC 90/2/4/49.
75 Cardell 1912, p.325.
76 The memorial reads: 'To the memory of Robert Napier, Smith to Argyle, who died 17 March 1845 aged 85 years also Margaret McDonald [sic] his beloved spouse who died 15 April 1845 aged 75 years'.
77 Wilson and Reader 1958, p.45.
78 Patent 7343, 1837.
79 Patent 1740, 1853. The exception was for 'improvements in machines for planing and slotting' (Patent 150, 1858) which was taken out in the joint names of James Murdoch Napier and William Thorburn, presumably an employee.
80 Wilson and Reader 1958, pp.43-44.
81 Patents 2523, 1857; 3113, 1857; 2438, 1859.
82 Patents 508, 1855; 1281, 1862.
83 Patents 555, 1855; 2208, 1857; 2247, 1860.
84 1861 census.
85 McDonald 1965, p.34.
86 1871 census.
87 Wilson and Reader 1958, p.46.
88 For an account of Robert Napier's business affairs see Hume and Moss 1979, pp.23-29, 49-50.
89 Cardell 1912, p.267.
90 The source for much of the early history of the firm in Wilson and Reader 1958, although they omitted some interesting details such as Napier's involvement in the Earl of Rosse's telescope.
91 Cardell 1912, pp.303, 306.
92 Cardell 1912, p.303.
93 Wilson and Reader 1958, pp.49-50.
94 Cardell 1912, pp.323-24.
95 Cardell 1912, p.307.
96 Patents 1447, 1867; 1552, 1857; 2595, 1867; 467, 1871; 1705, 1871; 402, 1872; 2639, 1872.

6
Joseph Clement
A.P. Woolrich

Joseph Clement had an enduring influence upon machine tool design and screw cutting.[1] He achieved this despite having only a basic education and no formal training as an engineer, rising to become Bramah's draughtsman and works superintendent, and subsequently Maudslay's chief draughtsman and marine engine designer. He did much to formalise the shapes of tools used in the fixed tool-holders of machine tools, where they were controlled by screw operated slides, and the forms he adopted have been used in general engineering ever since.[2] Clement's reputation has been eclipsed by those of Maudslay's more famous pupils and workmen such as Nasmyth, Roberts, and Whitworth, for he remained a working craftsman after leaving Maudslay, never employing more than a handful of staff. If he is now known at all, it has been for the tensions in his relationship with Charles Babbage, with an implication of dishonesty on Clement's part, and this has largely buried the proper investigation of Clement the engineer and craftsmen, and his importance in the history of machine tools.

Clement was baptised at Great Asby, near Appleby, Westmorland, on 13 June 1779. His father, Thomas Clemmet, was a weaver, much interested in nature, and an amateur mechanic. After education at the village school, Joseph Clement worked at his father's trade before becoming a slater. He learnt metal-working with a local blacksmith in his spare time, and built him a lathe. Clement became a proficient turner, making himself a wooden flute and his father a microscope to assist in his nature studies. He also learnt screw-cutting, made himself a set of die-stocks in 1804 and by 1805 was making power looms at Kirkby Stephen, working for a Mr George Dickinson. He was subsequently employed by Forster & Sons of Carlisle to make looms.[3] In 1807 he had moved to Glasgow, where Peter Nicholson, a well-known writer on carpentry, was able to assist Clement in his efforts to learn engineering draughtsmanship.[4] Clement then worked in Aberdeen for Messrs Leys, Masson & Co., designing, making and fitting up power looms. While at Aberdeen he made a lathe capable of screw-cutting, fitted with a small slide rest. During 1812-13 he attended a course on Natural Philosophy taught by Prof. Copland at the Marischal College, Aberdeen.[5]

In 1813, aged thirty-four and with £100 saved, Clement moved to London where he first took a job with Alexander Galloway at a guinea a week, and soon after with Joseph Bramah where he became works manager and chief draughtsman, tripling his wage. Following Bramah's death, the works were taken over by his sons who resented Clement's influence and dispensed with his services, although his five-year contract from April 1814 had most of its term still to run. By 1815 he was chief draughtsman at Maudslay's where he worked on the design of the firm's celebrated marine engines.[6] Only eight years Maudslay's junior, Clement can in no sense be regarded as a pupil, like Nasmyth. Instead he was given a job with seniority, reflecting his increasing expertise. It is likely that the Maudslay drawing office absorbed new techniques from Clement, rather than the learning process being entirely in the other direction, as Roe and Rolt suggested.[7]

Clement left Maudslay in 1817 to set up his own business as a technical draughtsman and maker of precision machinery at 21 Prospect Place, Newington Butts, with a capital of £500.[8] By the late 1820s the workshop had expanded into a nearby chapel.[9] It has been suggested, although there is no clear evidence, that the business was financed by the Duke of Northumberland, an amateur mechanic who purchased a Maudslay lathe, and who was an admirer of Clement's abilities.[10]

Details of his professional life are sketchy, but Clement acquired a name as one of the best draughtsmen in London, making drawings for engineers and architects. Clement's skill in the production of high-quality machinery was second to none, and he seems to have specialised in special purpose machines and tools, and made improvements in lathe design. His business remained relatively small, employing thirty men at most, though these were 'of the first class' and Clement's example of care and accuracy gave his works a reputation as a fine training ground for mechanics.[11] All this was achieved despite his own lack of social graces; Clement, heavy-browed and unpolished in manner and speech, retained his strong Westmoreland dialect. Smiles claimed that 'he read but little, and could write with difficulty', but this contention is not supported by Clement's very literate accounts of his inventions in the *Journal of the Society of Arts*, nor by letters in the Babbage collection relating to his work on the difference engine.[12] Unpolished he may have been, but Clement was astute enough to manage men successfully, to run a craftsman's workshop, and to be able to conceive, design and manufacture special purpose machinery.

So far as is known, none of Clement's original drawings have survived, other than some undertaken for Babbage.[13] Of these, it is not clear how many are in Clement's own hand, or whether they were by his draughtsmen. From 1817, Clement illustrated articles in the *Transactions of the Society of Arts*, and later in Thomas Tredgold's *Treatise on the Steam Engine* (1827); although the drawings were necessarily modified by the engraver, concealing the skill of Clement as the originator. Smiles claimed that in some of his best drawings Clement 'reached a degree of truth in mechanical perspective which has never been surpassed'; but this is an exaggeration as contemporaries such as John Farey junior (1791-1851) had developed to a high degree the skill of producing accurate perspective drawings of machinery.[14]

In 1818 Clement was awarded a gold medal by the Society of Arts for his instrument for drawing ellipses. His nomination for this honour was endorsed by the eminent engraver Wilson Lowry, famed for his engravings of machinery in publications such as Rees's *Cyclopaedia* (1802-1819).[15] The machine for drawing ellipses was vital for engineering and architectural engravers, as it automatically generated accurate ellipses and curves on the copper plates from which the drawings were printed, a far higher degree of accuracy than could be achieved by hand. While empirical methods of producing ellipses and curves had long been known to the draughtsman, these were not accurate enough for machinery representation for workshop use, for patent specifications, or for the increasing numbers of technical books and periodicals then published. Engravers of machinery drawings also needed to be able to cut curved lines accurately on the plates, as manual execution was very liable to error. Modified ellipse machines were specially made using a hard point, perhaps made of diamond, to cut the metal instead of the drawing pen or pencil of the draughtsman. Wilson Lowry had one for making the repeated rulings used to indicate shadow and texture. Clement's ellipse machine had the feature of being fitted with a division plate so that ellipses

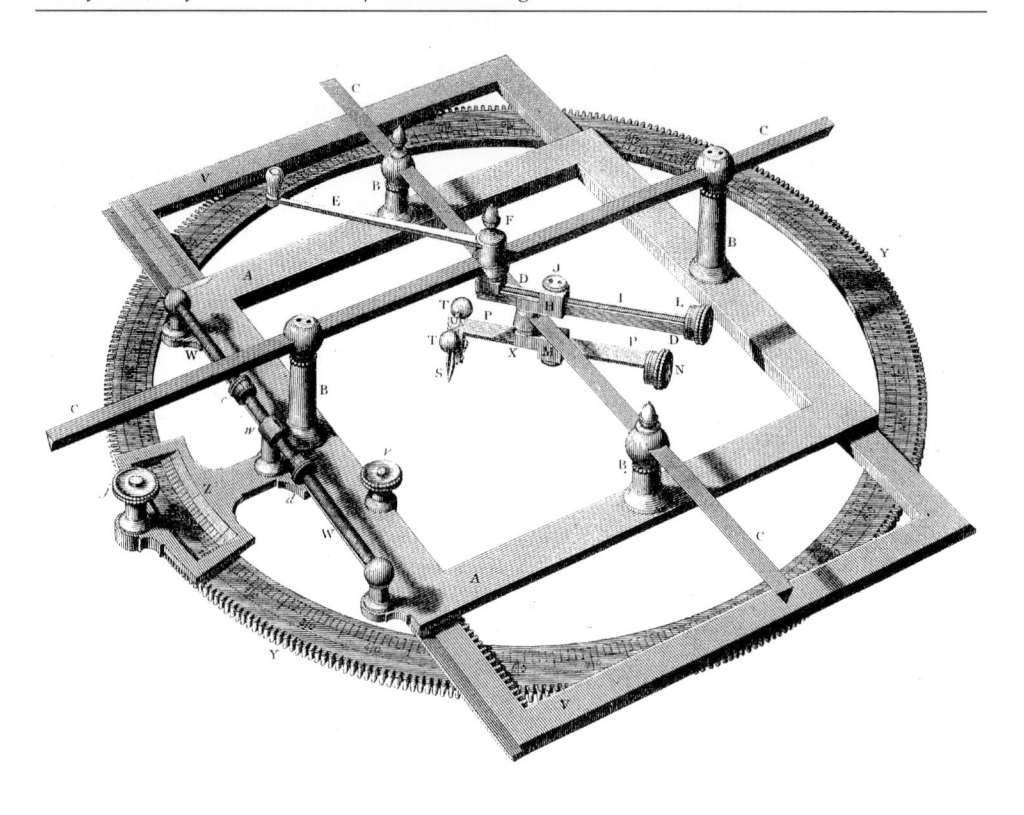

Fig.46, above: Clement's instrument for drawing ellipses. *(TSA)*

Fig.47, below: An example of the use of Clement's instrument for drawing ellipses. *(TSA)*

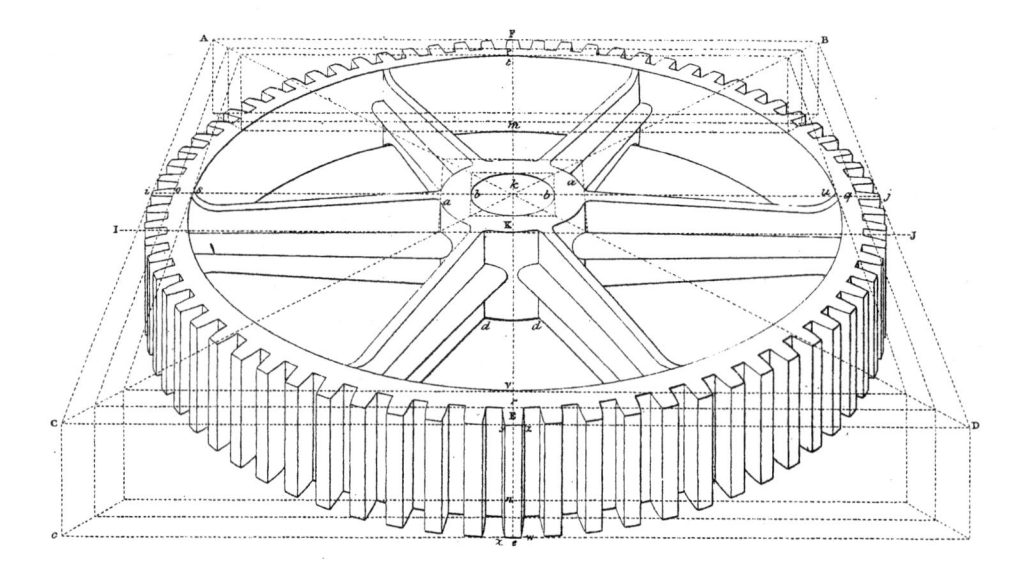

could be accurately divided, therefore making possible perspective drawings of objects such as gear wheels.

Clement made a further contribution towards drawing office technology in the form of a drawing table or stand for large-size drawing boards, details of which were published in the *Transactions of the Society of Arts* for 1825.[16] He had been asked to make some drawings 7ft by 6ft in size, impossible to execute on a traditional board without climbing on to reach the centre. In design, Clement's table was similar to the present day draughtsman's easel stand, allowing the board to be used vertically or horizontally and at all angles between. In addition, the board itself could swivel on a pivot-pin, allowing the drawing to be worked on from all sides, as most convenient to the draughtsman, and would also slide up and down on runners. These features allowed universal access and the stand accommodated boards of different sizes. Clement's practice was to hang the tee-square at the top of the board and rule lines perpendicularly. If lines were needed at right-angles to these he would swivel the board through ninety degrees and repeat the operation.

As an engineer, Clement was famous in his lifetime for his planing machine and his lathes. He was not the first man to make planers, for they were also produced by James Fox of Derby and Matthew Murray of Leeds. It was the planing machine in particular which permitted the transition in machine design to all-metal construction. Before it, most machines, especially heavy ones like the steam engine, were framed in wood, to which a minimum number of metal parts, usually blacksmith-forged, were fixed by nuts and bolts. Machines made in this way needed constant attention to make them work well, since the wood racked with humidity and temperature, causing the metal parts to lose adjustment very easily.

Fig.48 Clement's drawing table for large mechanical and architectural drawings. *(TSA)*

Clement originally made a planer in 1820 for machining the sides of triangular lathe beds, and for loom components. A few years later he made his great planer, from working drawings, without constructing a model. It was never patented, as Clement relied upon his own and his workmen's skill in using it as protection against competition.[17] The machine comprised a bed running on rollers, to which the casting to be machined was fixed, which passed to and fro beneath a fixed tool. Clement's planer had a cast iron bed, level with the floor, 19ft long, 4½ft wide, and 6in deep. It was anchored to a brick platform 3ft deep. Underneath this was a stone 6in thick and 5ft square, to which the frame holding the tool was anchored with through-bolts. The interstices of the iron bed were filled with cement, binding it to the brickwork, which had the effect of helping to prevent spring in the metal and so damping down vibration in the tool cut. To this iron bed were fixed a number of iron wheels, running on accurately machined bearings. On these wheels ran another iron bed to which the work to be machined was fixed. Great care was taken to ensure the absolute accuracy of the levels of the wheels. Under the moving bed were fixed runners sitting on the rollers and a pair of racks engaging in pinions operated by winch-handles. When the operative turned the handles, the bed moved backwards and forwards past the tool, removing a shaving of metal at each pass. The bed was equipped with spring fittings to dampen the shock at the end of each stroke. The machine was also fitted with two tools, set facing opposite each other, so a cut was made on each stroke, the tools being disengaged and reactivated in turn.[18] The planer could be fitted with centres and used for planing circular, spiral and conical, as well as flat work. The cutting speed was evidently low, since one man could keep it in motion for ordinary jobs, though two were necessary for making long and full cuts both ways. For more than ten years it was the only one of its size in the London area and ran for many years, day and night, on jobbing work, forming Clement's primary source of income. Smiles notes that his charge for planing was 18s per square foot, which earned him £10 for a twelve-hour day.[19]

Clement also made lathes, though whether these were for general sale is not known; more likely they were for his own use. He was awarded the Isis medal of the Society of Arts in 1832 for a lathe which permitted the speed of the metal past the cut to vary automatically with the diameter.[20] At that time the steels used for metalworking tools could vary greatly in quality,

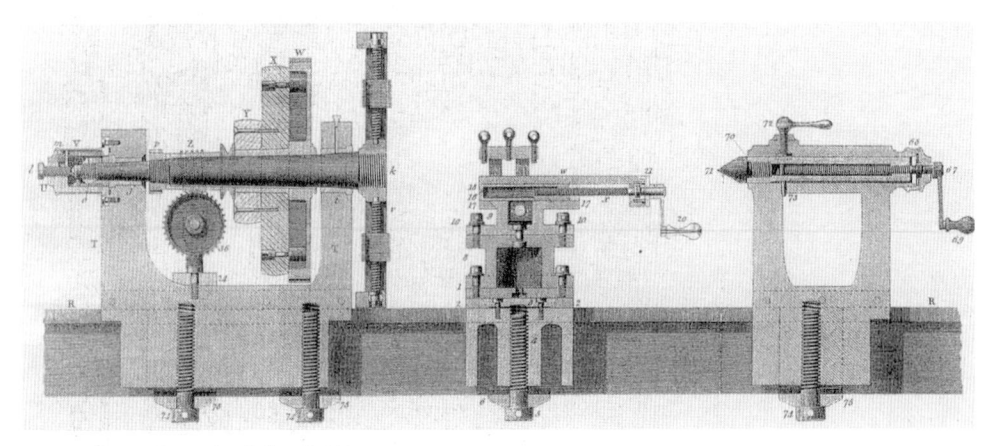

Fig.49 Clement's turning lathe. *(TSA)*

Fig.50, above: Clement's planing machine (side elevation). *(TSA)*

Fig.51, right: Clement's planing machine (end elevation). *(TSA)*

some softening very rapidly if overheated. In his 1829 paper delivered to the Society of Arts, Clement quoted a linear speed for cast iron of 15ft a minute. This reflected the need for a slow speed to cut beneath the hard silica skin left by the moulding sand on the iron's surface, and also the soft tool steels then used.[21] The problem of varying speeds can be acute in facing a disc in the slide rest lathe, where if it revolves at constant rotational speed, the linear speed past the tool tip increases in proportion to the distance of the tool from the centre of the lathe. A skilled turner can vary the rate at which the tool moves in and out from the centre line of the lathe, keeping the linear cutting speed constant, but for an unskilled turner, keeping the tool moving at the same rate is hard, with the result that tool tips burn out and the work being turned is often spoiled. Clement overcame this problem by building an elaborate lathe with a complex mechanism which ensured consistency of motion wherever the tool tip was in relation to its distance from the centre line of the lathe. The lathe was hand-driven by belting through a pair of opposed cone riggers, permitting infinitely variable rotational speed of the chuck, restricted only by the effort put in by the person who turned the handle powering the lathe. The lathe drove a mechanism which could detect where the tool point was in relation to the centre of the lathe, which in turn could vary the position of the belts on the cone pulleys, so ensuring constant linear speed past the tool tip. It was possible to set up the lathe with different gear and belt ratios for machining the various materials – wrought iron, steel, cast iron, brass, and so on – then in use. The lathe was fitted with micrometer adjustments on the two motions of the top and cross slides. It was not equipped for screw-cutting[22] and seems to have been simply a mechanism for facing work held in the chuck.

While Clement's lathe was an awesome piece of design, it was basically impractical. It must have been a nightmare to keep in good order, with its proliferation of flat and round belts, gears and clutches. It seems to have been a special purpose tool developed for use in his own workshop and certainly had no influence on lathe development. Clement is also known to have had a more normal pattern screw-cutting lathe, but few details have come down to us. The bills he presented for work on the difference engine in the early 1830s mention turning components in excess of 7ft long, so we must assume his lathe could accept work of at least 8ft in length.

Clement did make two contributions to lathe design which have stood the test of time. The first was a double driver, known as a Clement driver. He devised it to enable work to be turned absolutely accurately. When turning work between centres, it is gripped by a carrier, the peg of which is driven by a pin on the catch plate screwed to the lathe mandrel. The effect of having a single point of contact was to bend the work at the point it met the tool tip, a result of the twisting action of the force applied. Clement overcame this by designing a double driver with two pins, which equalised the forces tending to make the lathe cut eccentrically.

His second contribution was the barrel tail-stock. In many lathes of that time the back-centre was screwed on the outside and turned by a winch handle to advance the tip of the centre into the seating cut into the end of the work being turned. As it was advanced it rotated, and this meant that the back-centre tip would not always line up accurately on the centre line of the lathe through the head and tail-stock. Clement overcame this by inventing the barrel tail-stock. The action of turning the screw by the winch handle pushed to and fro the barrel, on the end of which was screwed the back-centre. The barrel was prevented from rotating by means of a small pin sliding in a groove machined along the length of the barrel.

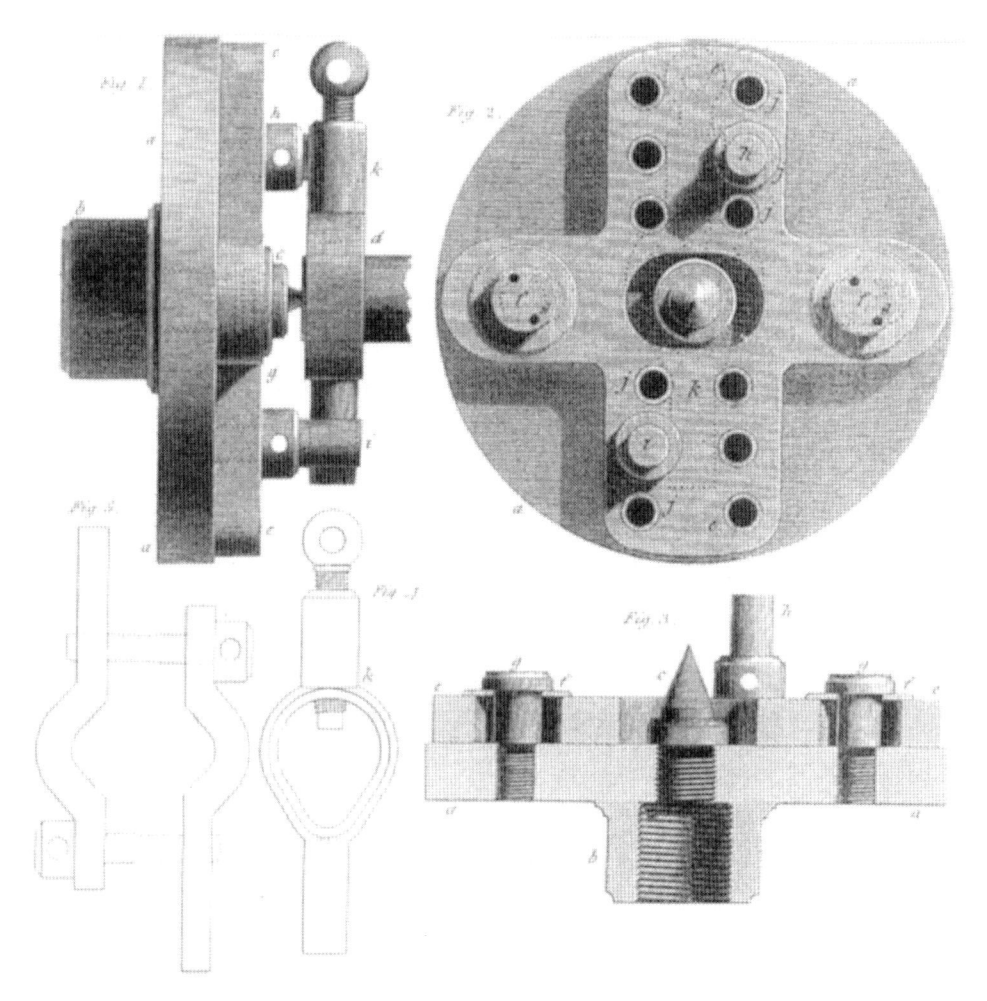

Fig.52 Clement's driver. *(TSA)*

Clement's design remains substantially unchanged in present day use, although taper-shanked back-centres, which are gripped by the barrel and prevented from rotating, have since been introduced. Likewise taper-shanked drills and reamers can be held for use. Taper fitting depends on the friction between the mating parts to lock them together against rotation. The back-centre is inserted sharply into the tail-stock barrel, and this is enough to make it grip.

Clement made an important contribution to the techniques of screw-cutting, and clearly gained much from the pioneering work of Maudslay between 1799 and 1810 on the provision of taps and dies for workshop use. Maudslay had developed the design of die which cut the screw thread, introducing three cutting edges in taps and dies, and dividing taps into three – the entering or taper tap, the middle tap, and the plug tap – enabling a thread to be made to the bottom of a blind hole. Maudslay made a series of taps ranging in size from 6in down to the finest screws used for watch work, with varying pitches from three threads per inch downwards. Clement later developed his own range of taps and dies, between three

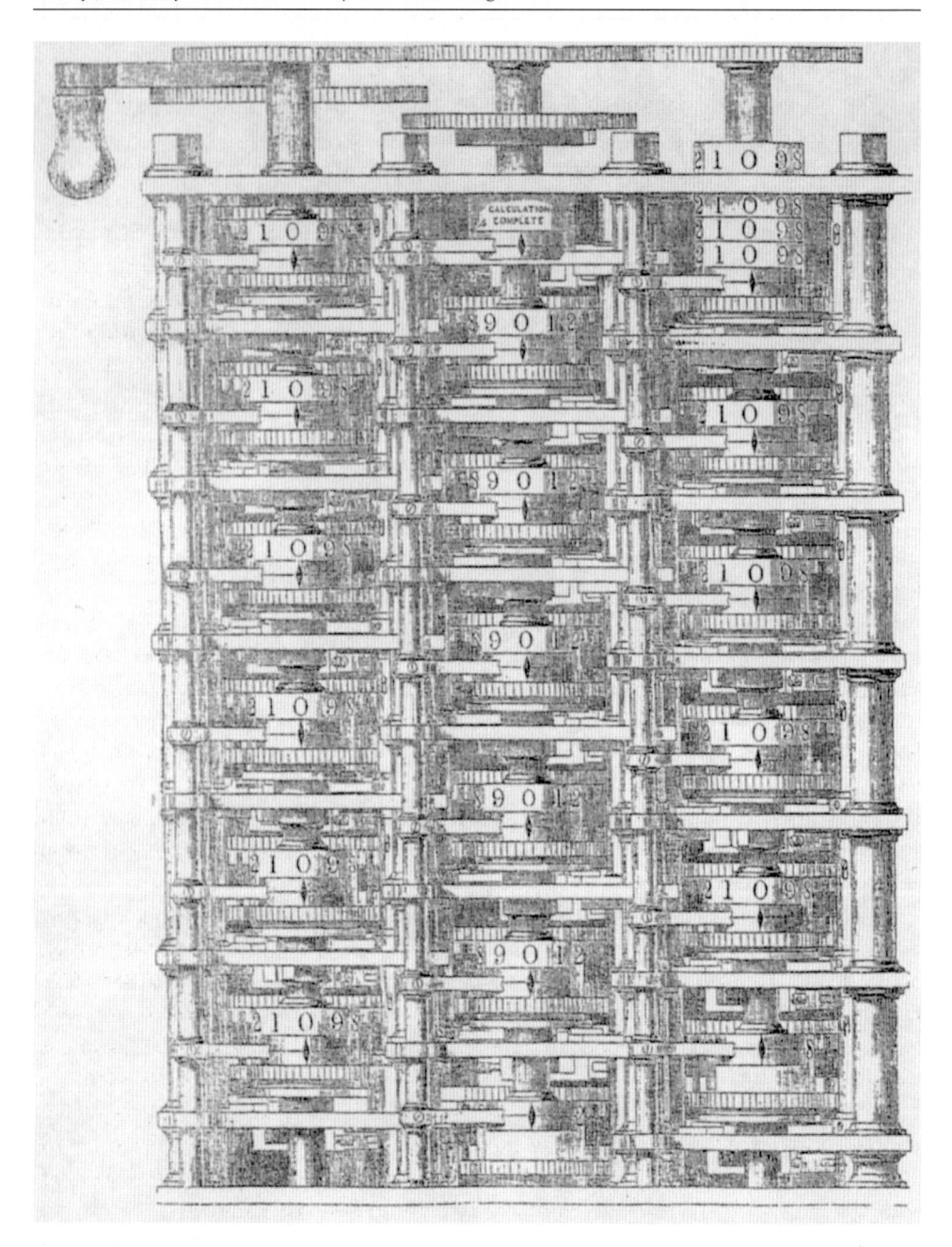

Fig.53 Babbage's Difference Engine. *(Babbage, 1864)*

quarters and 1½in in diameter and all with a pitch of seven threads per inch. It is not known whether he developed a smaller range of sizes; he may have used the kits of screwing tackle made and sold to the trade by Holtzapffel & Co.[23]

Clement is credited by Smiles with the use of the milling cutter to cut the flutes of taps. The introduction and use of the milling cutter in engineering production has not been well researched, but it is known that small rotary gear-cutters were in use by clockmakers at the beginning of the nineteenth century, and were also used for machining small metal parts in the lathe where the work was gripped by the tool-holder. A tool-grinding machine for sharpening clockmakers' inserted tooth-milling cutters is illustrated in Rees's *Cyclopaedia*.[24] Clement is also credited with the 'headless' tap, allowing a nut to be lifted off over the tap's shank once it had been threaded. In this way a number of nuts could be finished before the filled shank was removed from the tap wrench.

According to Smiles, by 1826, recommended by Marc Isambard Brunel, Clement was engaged by Charles Babbage to manufacture the parts for the Difference Engine, antecedent of the modern day computer.[25] The relationship with Babbage lasted for about six years, latterly deteriorating because of tensions between the two over managerial accountability and financial control.

Charles Babbage devised and built a number of mechanical calculating machines with the intention of printing mathematical tables used in calculating mortality and annuities, and for astronomical and navigational purposes. Babbage's concept was to have numbers punched on copper plates to produce the tables required. These plates were then used to make stereotypes for the printing machine. Being automatically calculated, the chance of error (which had bedevilled tables made by human calculators) was eliminated. Babbage began working on his Difference Engine No.1 around 1823, and employed a small number of workmen in his house on development work. Once a full-size machine became feasible, a professional craftsman was essential.

Financial records suggest that Clement became involved in building the Difference Engine No.1 in 1826, though he had possibly made a special machine tool for Babbage earlier, during 1825.[26] He worked on the project for two blocks of time, from 1826 to May 1829, and from spring or summer 1830 to March 1833. Clement had no involvement with Babbage's later machines, the Difference Engine No.2 and the putative Analytical Engine. A curious feature of the records is that Clement was not mentioned by name in any of Babbage's papers until 1829, three years after he started work on the project.[27] While Babbage kept even the most trivial note from acquaintances asking to see the machine at its various stages of development, he did not apparently retain a scrap of correspondence regarding its design and construction. Most of the records are financial, and these very largely relate to his later dispute with Clement. While it is possible that all the documentation was housed in Clement's workshop, which Babbage visited very frequently, it does not explain why Clement was never mentioned by name in Babbage's letters to others. It is as though he felt that Clement was not important enough to be recognised – as indeed Babbage showed later in his book *Passages from the Life of a Philosopher* (1864), where Clement's name does not appear and he is noted dismissively only as 'The Engineer'.[28] The imperfect survival of records, and the partial nature of Babbage's version of events, mean that a full understanding of the relationship between Clement and Babbage is not possible.

Babbage's Difference Engine stood 9ft high, and was 9ft long by 3ft 3in deep. Seen from the front it was 'L' shaped, with a calculating division on the left, a printing division on the right, a part to link the two, and framing and chassis running on rollers. The main material used in the construction was gun metal, with steel shafts. By the time the project foundered in 1833, some 12,000 parts, weighing four tons, had been made by Clement. The project was funded by the Civil Contingency Fund of the Treasury, which shielded the Government and Babbage from Parliamentary scrutiny, and by the time of its abandonment had cost the Government £15,288, of which the bulk was Clement's charges.[29]

The accounts provide an insight into Clement's workshop processes, and raise questions about the machines with which he was equipped. Numerous components are listed – calculating wheels, figure wheels, sector wheels, and various clutches – and it can be assumed that Clement had a furnace in his workshop for melting the metal. There were numerous gear-wheels, presumably cut with a gear-cutting engine, and components with complicated shapes, which in a time before the development of milling machines must have been hand-finished.

The Difference Engine was developed at a time of transition between craft working and mass-production, and before there were the means to make identical parts in quantity.[30] Components were produced individually, so that each was very slightly different. Although a skilled craftsman could, by careful filing, scraping and incremental machining, make parts which fitted and ran together, it was time-consuming and labour intensive – hence expensive. The lack of standardised screw threads, and even linear measurement standards throughout the industry, meant Babbage was unable to send out jobs to different workshops – parallel production as it is known today. The major mechanical engineers made their own sets of taps and dies, and every lathe had individual masters for screw cutting, each subtly different from the other. The use of the change wheel system for screw cutting did not become common until the end of the 1830s. The result of this was that a nut made in, say, Birmingham, to a particular size and screw pitch, was not guaranteed to fit a bolt made with the same dimensions in London. To ensure the most consistent standard, a machine had to be made in one workshop. Clement had his own range of screw thread pitches, and it is significant that he employed as a journeyman on the Difference Engine, Joseph Whitworth, who was to play a seminal role in the introduction of standards of measurement and of standardised screw threads throughout the engineering industry.

Soon after Clement started on the Difference Engine, a spell of ill health interrupted his work.[31] Clement was prone to long-term illness, sometimes bed-ridden for several months at a time, and was afflicted several times during the course of this project with something described by Babbage as a 'low fever'.[32] Babbage then suffered a series of personal tragedies in 1827: his father, son and wife died, leaving the widower with young children to raise. Babbage, distraught, hurriedly embarked on a tour of Europe lasting over a year, having left William Herschel to supervise work on the Difference Engine.[33] Babbage returned to England in November 1828, to find that criticism of his Difference Engine project had appeared in the press. It was claimed that the project had failed and that Babbage was making a personal profit from government funding. Herschel had defended Babbage, pointing out that he was in fact out of pocket.[34]

Babbage left funds in Coutts's Bank for Herschel's use on the project, but little is known about day-to-day progress during Babbage's absence.[35] He later claimed to have left drawings and instructions sufficient for the work to proceed. It is difficult to see how Clement can have dealt with the design problems and difficulties which would have inevitably arisen, apart from using his own judgement and hoping that Babbage would concur on his return. Babbage's letters are completely silent about Clement until 11 April 1829, when he was negotiating with the government to ascertain the extent of their commitment to the project. To Bryan Donkin, Babbage wrote that he had concealed from Clement the government's involvement, since he thought that if he presented himself as a private individual, Clement's charges would be more moderate.[36] Clement of course knew the situation, from the previous year's press correspondence,[37] which may explain his apparent hardening of attitude towards Babbage.

Babbage's approach to the project had been far from businesslike. He had been told in June 1823 by the Chancellor of the Exchequer, John Robinson, that the Government would fund the project, and was given a Treasury Warrant for £1,500. No budget or price ceiling had been agreed, nor a plan of production. By 1829 the Treasury was expressing grave doubts, so Babbage wrote to the Prime Minister, the Duke of Wellington, pointing out that he had already spent around £6,000, of which £4,500 had come from his own pocket. Babbage received a further £1,500 in April 1829.

The rocky financial status of the project caused Clement to become uneasy about its viability. He had, after all, invested a great deal of his time and skill, and was responsible to Babbage for paying the wages of the men,[38] so it was a matter of concern if the client was unable or unwilling to meet his obligations. In turn Babbage became unhappy at what he regarded as imprecise invoices from Clement. Babbage obtained from Clement an agreement to submit his bills to arbitration, which later found Clement's accounts to be perfectly fair. Clement nominated Henry Maudslay to represent him as assessor in the arbitration; following Maudslay's death, his partner Joshua Field took on the task. Babbage's nominee was Bryan Donkin.[39]

Babbage and his friends appear to have admired Clement's work while questioning his charges. Babbage told Bryan Donkin, one of the arbitrators, in April 1829, that Clement was 'a most excellent workman and draftsman and ought to be well paid'. But in May that year, Richard Penn, who had helped set up the arbitration, referred to Clement's 'offensive conduct' and 'low cunning', and suggested that Babbage 'force the matter to a more hostile arbitration'.[40] The Duke of Somerset wrote to Babbage that Mark Isambard Brunel had spoken with admiration of the accuracy of Clement's workmanship, which surpassed everything he had seen. While the Duke was still 'very angry' with Clement, Brunel had recommended reaching a compromise.[41] Isambard Kingdom Brunel was also to become involved in a dispute over Clement's charges for a hundred steam whistles for Great Western Railway locomotives, which the company's works were unable to produce to standard. Brunel claimed that the bill was six times more than the whistles were worth, but lost the subsequent arbitration.[42]

A confirmation of Clement's importance to Babbage comes in a letter written in November 1829 to Lord Ashley, where Babbage proposed that a continuing system of arbitration be established to check Clement's charges. He again described Clement as 'a most excellent draftsman and an able workman'.

> He has been constructing the machine for me and into his head I have for several years been conveying all my ideas on the subject of the machine and he is consequently in full possession

of them. At several periods during this interval he has been so ill as to be in a state of the greatest danger and I cannot describe to you the anxiety with which in such circumstances I have on coming within sight of his house strained my eyes to see if the windows were closed; and I, by his death, deprived of the result of years of anxious labor. Much of that labor is now fixed in drawings which it would require much time to make another person fully understand and much of the machinery is already executed still much remains in his mind ready to be produced and it is clearly of the greatest advantage to the progress of the machine that this should continue under my direction to execute it.

Babbage complained, though, that since Clement had found out about the government's financial assistance, 'I have found greater difficulties in my communication with him; and I am much displeased with several parts of his conduct'. Babbage had asked the engineers John Rennie and Bryan Donkin to judge Clement's charges, in the light of Babbage's opinion that 'such labors attended with such loss of health as he had experienced deserved to be well remunerated'. Babbage suggested that he had taken this course to avoid a direct row with Clement over the bill, in order that work could meanwhile continue on developing the machine. Any discussion of the costs 'might possibly have betrayed him into irritation of temper a state very unfavourable to his receiving my instructions relative to the machine'. Babbage wanted to avoid future disputes by having the government take over ownership of the machine, and payment to Clement, while he himself would retain direction of Clement's work.[43]

This phase of the dispute was settled and work recommenced in May 1830, with an estimated further three years of work to be needed. Work progressed for about two years, and Babbage began to look towards assembling the machine from the manufactured parts. He acquired premises near his house, where a fireproof workshop was to be built. It was envisaged that Clement would live on the premises, but since his operation would need to be split between the old and new workshops, he claimed compensation of £660 per year. The Treasury rejected this, and the relationship between Clement and Babbage entered its death throes. After a succession of demands and threats, with both men becoming more and more intransigent, the workmen, including Joseph Whitworth, were laid off at the end of March 1833 and Clement's involvement ceased. It took a further sixteen months of complex negotiations involving third parties and arbitrators before the drawings and parts were finally delivered to Babbage, and Clement received what he felt was due payment.[44]

Between 1833 and his death eleven years later, Clement continued business as a workshop engineer, relying on his reputation for accuracy and precision to attract orders. Smiles tells us that Clement's inventive faculties were always stimulated by what he regarded as a 'tough job', and one of the engineer's last works was the construction of a pipe organ. Clement died of apoplexy on 28 February 1844 at 31 St George's Road. The death was reported by his nephew, Joseph Wilkinson, who recorded the surname as Clemmet, his baptismal name. Clement never married, though he fathered an illegitimate daughter, named Sarah, by Agnes Elson. The circumstances surrounding this liaison remain obscure. At the time that Clement made his will in 1839, mother and daughter were living in Tanfield near Aberdeen. In his will, Clement set aside £4,000 for his 'natural daughter', together with his gold watch chain and seals, and the Society of Arts gold medal which he had received for the instrument for drawing ellipses. Clement's nephew, Joseph Wilkinson, was given the option of buying his uncle's machinery and tools of trade for a further £4,000. It appears that this option was taken up, as Joseph Clement's business was continued by his nephew at the Newington Butts workshop. Clement clearly had a strong sense of family duty, for he also made provision for his four siblings, their spouses and children, together with various friends living in and around the place of his birth.[45]

Clement's keenest admirers were almost certainly fellow engineers, for he never patented an invention, neither did he produce a machine likely to capture the popular imagination. His most important work was arguably the development of Maudslay's ideas on screw-cutting, and his encouragement of Whitworth along a similar path. Two years before his own death Babbage noted how it was 'frequently said that: Mr Babbage made Clement. Clement made Whitworth. Whitworth made the tools'.[46] The perfection of screw-cutting and screw standardization was most probably a collective achievement, and one in which Clement was a key player. Nasmyth positioned Clement after Samuel Bentham and Maudslay in the hierarchy of early engineers and claimed that his 'admirable mechanical contrivances and tools', executed in connection with the calculating machine, were a spur towards 'the perfect in mechanism'.[47] William Fairbairn placed Clement alongside Maudslay, Murray and Fox as one of the 'great pioneers of machine-tool making'.[48] Certainly Clement was at the very centre of improvements in machine tool design during the opening decades of the nineteenth century.

Endnotes

[1] There is little reliable information about the life of Joseph Clement. Samuel Smiles's account in *Industrial Biography* 1863, pp.236-57, informed later writers on machine-tool history, such as Roe 1916 and Rolt 1965, both of whom quarried Smiles extensively. Smiles gained his information about Clement from John Penn (1805-1878), the Greenwich engineer, and from Joseph Wilkinson, Clement's nephew, who continued the business after his death. Nothing is known about Penn's link with Clement.

[2] Holtzapffel 1846, 2, pp.531-35, 537.

[3] Smiles 1863, pp.236-39.

[4] Smiles 1863, pp.239-40.

[5] Smiles 1863, p.240. Patrick Copland (1749-1822) enjoyed a considerable reputation as a teacher, set up a noted museum of natural philosophy, and was among the first to extend the knowledge of science beyond academic circles through his popular course of natural philosophy; see *DNB*.

[6] Smiles 1863, p.243.

[7] Roe 1916, p.46. Rolt 1957, p.12.

[8] This address, near the Elephant and Castle, later became 18 St George's Road.

[9] Swade 2000, p.70.

[10] Smiles 1863, pp.243-44. I am grateful to Michael Wright for ideas on this point.

[11] Smiles 1863, p.257. Men who had been employed on the Difference Engine had 'a greater chance of the best work' and were able to command higher wages than other workmen. R. Wright to C. Babbage, 18 June 1834: BL, Add. Mss. 37188, f.390.

[12] Smiles 1863, p.257.

[13] Now in the Science Museum Library.

[14] Smiles 1863, p.244. Clement made the drawing of Napier's tracing machine that appeared in the

TSA, see Fig.39.

[15] *TSA*, 36, 1818, pp.133-77. John Farey junior was awarded a gold medal for a similar device, called an elliptograph: Woolrich 1998.

[16] *TSA*, 43, 1825, pp.138-42, plate 11.

[17] Smiles 1863, p.249.

[18] *TSA*, 49, 1832, pp.157-85.

[19] Smiles 1863, pp.249-51; Roe 1916, p.52.

[20] *TSA*, 46, 1829, pp.67-105.

[21] *TSA*, 46, 1829, p.70. By the end of the nineteenth century, improved steels had increased this to 40ft a minute, a decade later to 60ft a minute, and improvements to steel compositions during the twentieth century brought it to 80ft a minute: Lineham 1906, pp.142, 975; Sparey 1972, p.126.

[22] Contrary to Roe's assertion: Roe 1916, p.59.

[23] Holtzapffel 1846, 2, p.674, fn.†

[24] Rees's *Cyclopaedia*, 10(2), 2 July 1808. See entries for 'Cutter' and 'Cutting Engine'.

[25] Smiles 1863, p.254. For the development of the difference engine, see Babbage 1864; Hyman 1982; Swade 2000; Schaffer (http://www.wmin.ac.uk/media/schaffer/). Many of these accounts draw extensively on Smiles, but see also the original documentation on http://home.clara.net/mycetes/babbage/.

[26] See http://home.clara.net/mycetes/babbage/.

[27] R. Penn to C. Babbage, 4 April 1829: BL, Add. Mss. 37184, f.247.

[28] As C.J.D. Roberts has pointed out, Babbage did not mention his wife in the book so there may be nothing sinister in his treatment of Clement: personal communication, August 2001.

[29] W. Brougham to C. Babbage, n.d., but almost certainly April/May 1827: BL, Add. Mss. 37183, f.99.

[30] Swade 2000, p.45.

[31] W.W. Whitemore to Lord Goderich, 4 October 1827: BL, Add. Mss. 37184, f.396.

[32] Report No.1 on Babbage's Calculating Machine, 13 July 1830: BL, Add. Mss. 37185, f.264.

[33] Swade 2000, pp.52-54.

[34] Swade 2000, p.57.

[35] Among the Herschel letters in the Royal Society are several from Clement on this subject.

[36] C. Babbage to B. Donkin, 11 April 1829: BL, Add. Mss. 37184, f.252.

[37] Swade 2000, p.57.

[38] Clement employed about ten men on the Difference Engine and this work constituted 'the principal part of his business': J.F.W. Herschel to C. Babbage, 12 February 1828, Royal Society Herschel Collection 2.219 and J. Stewart to C. Babbage, 14 September 1832, BL, Add. Mss. 37187 f.134.

[39] Swade 2000, p.61.

[40] R. Penn to C. Babbage, n.d.: BL, Add. Mss. 37184, f.281.

[41] Duke of Somerset to C. Babbage, 8 June 1829: BL, Add. Mss. 27184, f.337.

[42] Smiles 1863, p.256. The date is not stated. Smiles notes a further instance of Clement's bills going to arbitration, that of a large screw for an American customer.

[43] C. Babbage to Lord Ashley, 25 November 1829: BL, Add. Mss. 37184, f.432.

[44] Swade 2000, pp.65-67.

[45] PRO, PROB 11/1994, quire 188.

[46] Swade 2000, p.70.

[47] J. Nasmyth to C. Babbage, 22 June 1855: BL, Add. Mss. 37196, f.251.

[48] Smiles 1863, p.299.

7
Joseph Whitworth
Angus Buchanan

Joseph Whitworth was a pivotal figure in the development of British engineering. He rose from obscurity with the huge successes of British industrialisation in the first half of the nineteenth century by demonstrating his talents as a machine-maker. By 1851, he had become one of the outstanding mechanical engineers in the world. His machine tools were of supreme excellence in their precision and construction. They were acclaimed at the Great Exhibition and in huge demand in engineering workshops throughout the industrialising world. But at the same time, he was among the first to recognise the deficiencies of British engineering in terms of general education and specialist knowledge and theory. He contributed significantly to tackling these deficiencies through his own philanthropy, and subsequent generations of mechanical engineers have had good cause to be grateful to him. Yet his intervention came too late to prevent the comparative decline of British engineering, because before the effects of improved educational practices could come into effect, engineers in Europe and the United States were able to seize the initiative. Even Whitworth's magnificent machine tools were made obsolete by the introduction of American standardised mass-production techniques, while French and German engineers displayed a confidence with thermodynamics and chemical theory which their British counterparts egregiously lacked. Whitworth went on to make brilliant contributions to armaments and to other branches of the engineering industry, but never with the complete domination of his earlier years. His career over the central decades of the nineteenth century thus displays a profoundly pivotal character, and it also demonstrates his determined attempts to influence the course of engineering in Britain.

Behind the monumental achievements of Whitworth's professional career there was a rather opaque and forbidding personality.[1] He was suspicious of the motives of others, evasive about his own origins, and inclined to be irritable and bad-tempered – characteristics that increased as he grew older. He did not welcome intimate confidences and was poor in personal relationships. He appears virtually to have abandoned his first wife. After visiting his workshops in 1846, Jane Carlyle described him as having 'a face not unlike a baboon' and speaking with a broad Lancashire accent, but she went on to admire his personality: 'one feels [oneself] to be talking with a real live man'.[2] Photographs of Whitworth in later life show a severe and well-bearded face and suggest a withdrawn temperament. Perhaps this unwelcoming personal aspect deterred potential biographers, as the recent study by Atkinson is, for all practical purposes, the first substantial work of this nature on Whitworth. Although generally sympathetic, Atkinson is by no means hagiographic in the treatment of his subject. But his text lacks historical detachment, and there remains a need to present Whitworth clearly in the context of his times, and to assess the significance of his distinctive training and of his outstanding contribution to British engineering. That is the objective of this chapter.

Fig.54 Sir Joseph Whitworth. *(The Whitworth Art Gallery, The University of Manchester)*

Whitworth was born on 21 December 1803 in Stockport, Cheshire. During his lifetime he was reluctant to talk about his early years except in terms of a success story for industrious self-help which he helped to promulgate. Atkinson has painstakingly pieced together an account of these years which is slightly surprising but almost certainly authentic. According to this account, Whitworth's father, Charles Whitworth, was a loom frame maker in the Stockport textile industry. When his wife Sarah died in 1814, he received an offer of training for the Congregationalist ministry. So he placed his three children in the care of foster parents and went off to pursue his vocation. On Atkinson's account, Joseph, who was eleven at the time, bore a grudge against his father for this desertion for the rest of his life, and never spoke to him again even though his father lived to 1870. Joseph's childhood was both hard and lonely, but he soon demonstrated an aptitude for working with machinery, which became his avenue to fame and fortune. Even though Stockport was a thriving centre for the textile industry at this time, the young Whitworth realised that to make progress as a mechanical engineer he would need to move to a larger centre of industrial activity. So by 1821 he had become a mechanic with Crighton & Co., a leading Manchester firm of textile machine-makers, where, he was proud to recall in later years, he earned his first journeyman's wages.[3]

Still pursuing his objective of acquiring the best possible training in his chosen trade, Whitworth set off four or five years later to seek employment in London. He was particularly anxious to be employed by Henry Maudslay in the Westminster Road workshop of Maudslay & Field. Quick to recognise his exceptional talent, Maudslay took him on and placed him alongside one of his most capable mechanics, even in that remarkable collection of engineering skills, the Yorkshireman John Hampson. Not much is known about Hampson, but there is a well-known anecdote of his response to Whitworth's success in achieving perfectly plane surfaces. Invited into Whitworth's home to see the result of his young colleague's labours, the taciturn Hampson exclaimed 'Aye! Tha's done it'. The need for such surfaces was well understood by Maudslay and his fellow mechanical engineers. They were essential to the development of precision engineering, and Maudslay had gone far towards establishing them by making well-ground planes available on the workbenches of his men. But perfection had eluded him, and it seems that Whitworth solved the problem by substituting scraping for grinding, and by preparing three matching plane surfaces simultaneously.[4]

It is necessary to preserve an element of reservation about any claim of originality on Whitworth's behalf in this matter. He was, after all, working as a junior craftsman under Maudslay's supervision at the time, and tackling a problem that was widely perceived to be a constraint on further progress. James Nasmyth, Whitworth's close contemporary who also served part of his training at Maudslays, had no doubt that full credit for creating perfectly plane metal surfaces was due to Maudslay. He wrote – with the assistance of Samuel Smiles – an articulate autobiography in which he described the achievement, without any reference to Whitworth. But although so close in training and business activities, Nasmyth and Whitworth became silent rivals who managed to avoid mentioning each other for most of their lives, so that Nasmyth's accuracy on this point is at least questionable.[5] It is also worth observing that Nasmyth arrived at Maudslay's workshop only in 1829, by which time Whitworth had already left, and Maudslay would have assimilated any improvements made by Whitworth into his workshop routine. What is certain is that ultra-high standards of mechanical engineering were established in Britain in the 1830s, that both Maudslay and Whitworth made important contributions to this achievement, and that it was Whitworth who went on to make these standards integral features of best engineering practice and to campaign tirelessly for their widespread adoption.

Some commentators have acknowledged Whitworth's qualities as a craftsman and mechanical perfectionist without being willing to accept him as an innovator. This judgement, however, is over-pedantic, considering that forty-eight British patents were issued in Whitworth's name. While many of these were concerned with points of mechanical detail rather than with completely new machines, they were often extremely important details that made the difference between a good machine and an excellent one. Moreover, in at least one instance – when he patented a machine for cleaning roads with a series of revolving brushes which swept the detritus up into a collecting tank – he invented a device which was both novel and exceptionally valuable in improving the quality of life in British towns.[6] These were the years when the young Friedrich Engels observed in lurid detail the conditions of life in Manchester at the same time as Whitworth was building up his enterprise there. Whitworth's contribution to improving the urban environment and public hygiene – a contribution that brought him the friendship of Edwin Chadwick and other social reformers – was thus of considerable social significance, even though he received little credit for it from Engels.[7]

Whitworth's qualities as an innovator were not confined to his mechanical inventions, but demonstrated themselves most eloquently in his skill for what later became known as 'production engineering' – that is, organising the processes of production in the most efficient manner. After leaving Maudslay in 1828, he had served briefly with two other seminal machine tool makers. First, he spent a year with Charles and John Jacob Holtzapffel, experts on lathe construction and operation.[8] Then he joined Joseph Clement and acquired a valuable insight into the workshop of that master craftsman – another product of the Maudslay 'school' – at a time when he was struggling to fulfil the demanding specifications for the construction of Charles Babbage's 'difference engine'. But Whitworth surmised correctly that Manchester, with its thriving cotton industry and heavy demand for machinery of all sorts, was becoming the new focal point for the engineering industry, and decided to return there. At the end of 1832 he set up his own workshop in a few rooms in Port Street in central Manchester, with a plate over the door bearing the legend: 'Joseph Whitworth, Toolmaker from London'. From the moment he returned to Manchester in 1832, he showed a rare talent for efficient workshop organisation. In 1833 he moved into larger premises in Chorlton Street, and it was here that he rapidly built up a model workshop devoted to the manufacture of machine tools. The products of this workshop were of an amazingly high standard of accuracy, and it was his success in maintaining this standard consistently that became Whitworth's greatest achievement, making his machine tools the best in the world.[9]

Fig.55 Whitworth's street cleaning machine. *(PIME, 1847)*

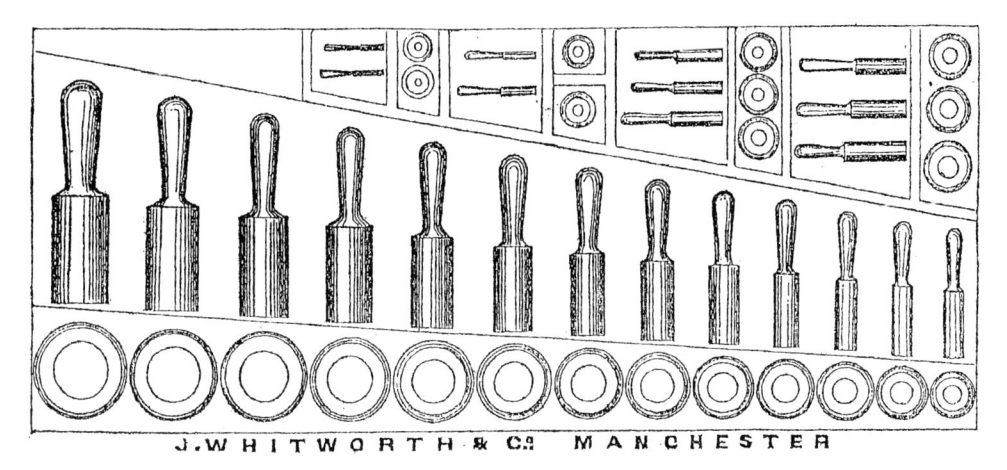

Fig.56 Whitworth's cylindrical gauges. *(1862 Exhibition Catalogue)*

Crucial to this consistency was Whitworth's emphasis on the use of distinctive tools and techniques: his plane surfaces; his gauges; his insistence on 'end' rather than 'line' measurement (that is, measuring by touch between points on the gauge, rather than by visual alignment as on a ruler); his superb measuring machine which was reputed to be able to register a millionth of an inch; his adoption of a precisely determined set of screw threads; and his use of decimal rather than fractional measurement. With the standard screw threads and decimalization, he became a tireless advocate for their adoption by the engineering profession, and such was the authority he derived from his huge practical and commercial success that both were assimilated into general engineering practice within Whitworth's lifetime. He took every opportunity to broadcast his views. In 1840 he spoke to the British Association, 'On Plane Metallic Surfaces, and the Proper Mode of Preparing them'. Then in 1841 he joined the Institution of Civil Engineers and promptly delivered a paper, 'On an Uniform System of Screw Threads'.[10] He subsequently joined the new Institution of Mechanical Engineers when this was formed in 1847, becoming President in 1856. The following year he addressed the Mechanicals 'On a Standard Decimal Measure of Length for Mechanical Engineering Work'.[11] He was still lecturing 'On Measurement' at South Kensington in 1876, and in 1881 the Board of Trade at last adopted Whitworth's standard gauges. The universal adoption of the standards of measurement advocated by Whitworth was a remarkable tribute both to the tenacity of the man and to the rationality of the cause.[12]

Manchester was a most congenial host to Whitworth's enterprise, and provided him with powerful encouragement in the development of his campaigns. In the middle decades of the nineteenth century the town was positively humming, not only with industrial activity and commercial success, but also with a rich intellectual life – shown in a wealth of stimulating societies and in support for 'progressive' movements such as those for free trade and urban improvement. Kargon has provided a vivid picture of this intellectual activity, distinguishing several stages in the development of scientific consciousness in Manchester. First, there were the amateur 'gentlemen of science' such as Thomas Percival and John Dalton, who established the Manchester Literary and Philosophical Institute around the turn of the nineteenth century. Then came the 'devotees' of science – men of substance who, while not themselves

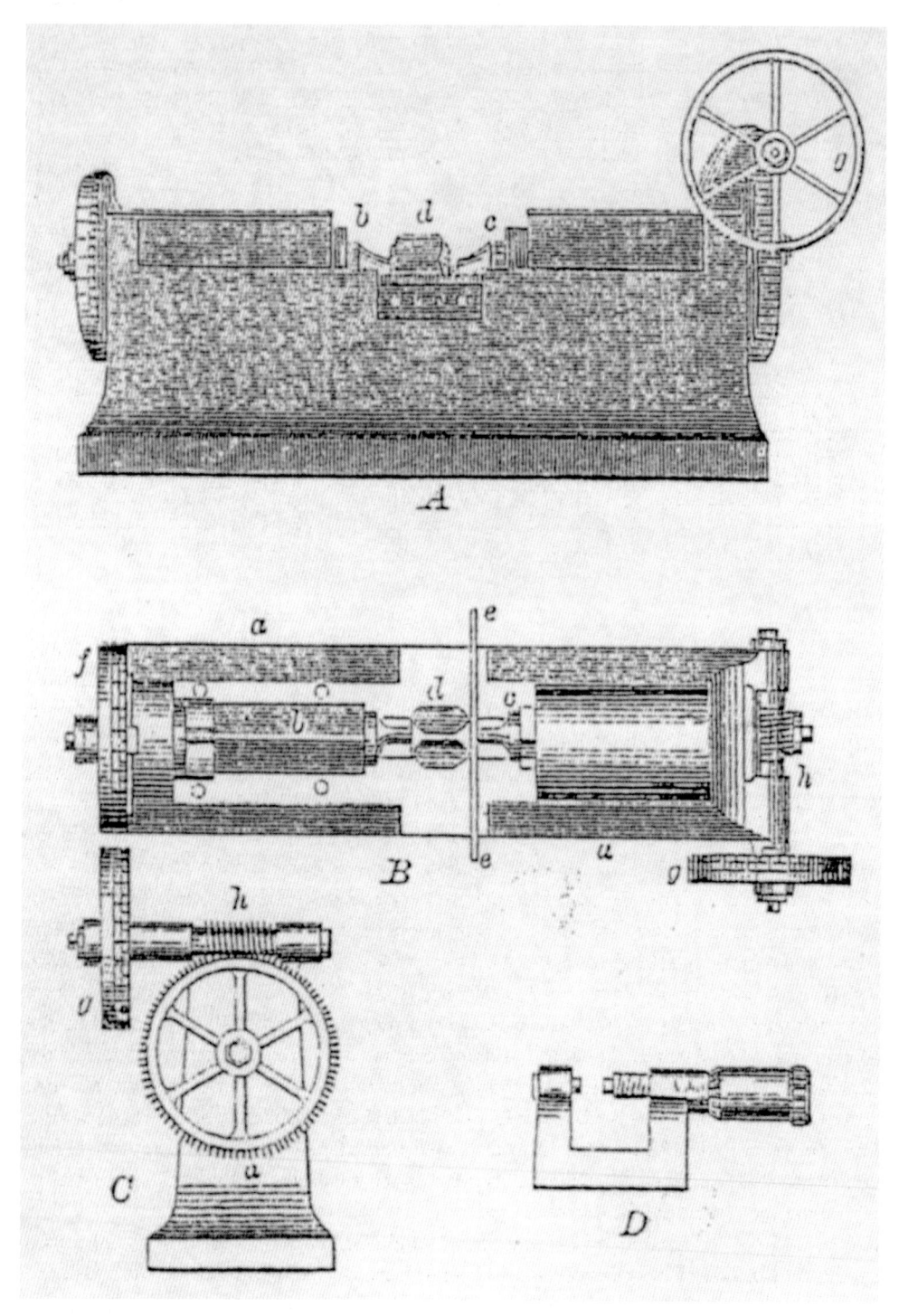

Fig.57 Whitworth's measuring machine. *(Knight, c.1870)*

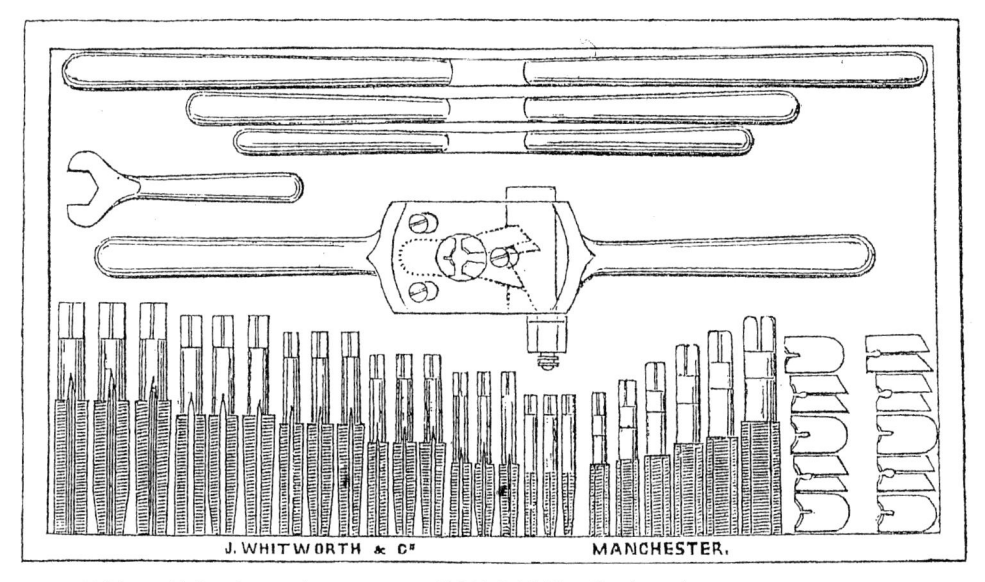

Fig.58 Whitworth's hand-screwing apparatus. *(1862 Exhibition Catalogue)*

scientists, exerted themselves to extend scientific institutions and representation. And thirdly came the 'professional' scientists, finding employment for their expertise in civic improvement and education. Kargon placed Whitworth firmly in the second stage, a 'devotee' who used his wealth and influence as an industrialist to promote scientific activity.[13] Whitworth found plenty of congenial company among industrialists like the Tootals and fellow engineers such as William Fairbairn, Charles Beyer, William Muir, and many other fellow devotees to the advance of science. Muir had followed him from Maudslay's and worked with him for some years before setting up his own business. He was a strict Sabbatarian and found Whitworth's relaxed attitude towards Sunday observance rather disconcerting. Fairbairn, in particular, became a close associate in organisations such as the Manchester Steam Users Association, formed in 1854 to tackle the problems associated with boiler explosions, with Fairbairn as President and Whitworth as Vice-President.[14] Whitworth seems never to have been tempted to embark on a political career himself, but he became friendly with the Manchester MPs Richard Cobden and John Bright, and turned to them frequently for support and advice. He entertained these and other friends regularly at the large house he had acquired at The Firs, Fallowfield. In conjunction with them he discussed plans for educational improvement in Manchester, and in time proved himself to be one of the most generous contributors to this cause. Manchester thus did much to shape Whitworth's career and to determine the character of his achievements.

Long before the full adoption of his uniform standards, Whitworth's success as a manufacturer had been assured by the excellence of his products. Between his return to Manchester in 1832 and the Great Exhibition of 1851, his firm underwent rapid expansion and his machine tools – lathes of all sorts, planing machines, and a wide variety of machines for cutting, milling, slotting and grinding – found their way into the workshops of the manufacturing nations. They were distinguished by their outstanding precision, simplicity of form, and absolute reliability. They enabled Whitworth to make a fortune as a machine manufacturer, but he was not content merely to get rich. He wanted the world to adopt the system of rational, precise, scientific manufacture that he had perfected, and he welcomed the Great Exhibition as an opportunity to impress the world with the potency of his message. He thus threw himself with enthusiasm into presenting his machines at the Exhibition. While most other contributors were content to show a single machine as representative of their work, twenty-three Whitworth machines were accepted including lathes, planing, shaping, drilling and boring machines, as well as his screw-cutting machines and his micrometer, which received particularly high praise. All were exhibited in Class 6, for manufacturing machines and tools. Whitworth's display caused quite a sensation and he won the prestigious Council Medal.[15]

After the Great Exhibition Whitworth became increasingly concerned about the education of engineers. The subject was one of long-standing interest to him. He appears to have received a sound basic education before the death of his mother. One of the few surviving business documents from his early career shows a neat hand and a careful writer. Even in old age, a rare manuscript letter shows that his handwriting remained clear and firm.[16] This in itself is remarkable, considering the lowly economic status of his family at a time when no state provision was made for primary education, and as an employer Whitworth must have been conscious of the fact that most of his fellow-workers lacked the advantages that literacy had given to him. He appears also to have been a keen reader. Atkinson has argued that Whitworth was a Baconian in his adoption of scientific empiricism, and there is some justification for this identification. But Whitworth was never an abstract philosopher, and even though he had been influenced by *New Atlantis* and other works of Francis Bacon in his youth, it is probably closer to reality to regard him as a thorough-going experimentalist by nature.[17] In any event, few of his colleagues shared his reading habits, and again it is reasonable to surmise that Whitworth's experience as the employer of a large labour force made him aware of the poor quality of education – if any – available to his workers. These anxieties were heightened both by the revelation of American enterprise and innovation at the Exhibition and by his experience of conflict with his own workforce immediately after.

The American evidence had to wait until the domestic conflict could be sorted out, because in the autumn of 1851, when the Crystal Palace had scarcely closed its doors, the British engineering industry faced a major disruption. This arose when the new Amalgamated Society of Engineers (ASE) presented some modest proposals about piecework and pay to the employers. This was a trade union conceived more as a friendly society, to provide various forms of social insurance to its members, rather than a set of rabble-rousing agitators such as those who had acquired an unsavoury reputation in the Chartist Campaigns of the previous decade. Sydney and Beatrice Webb later described them as representatives of 'New Model' trade unionism, and these moderate unionists were to have a profound influence on the development of the British labour movement. But perhaps their contemporaries among the employing classes could not have been expected to recognise the quality of this innovation, and they reacted in virtually unanimous alarm. The engineering employers of Manchester, in particular, combined to impose a lock-out designed to deny work to any of their skilled employees unless they rejected membership of the union. As the great majority of the best-trained workers had already enrolled this meant closing the factories until the union could be broken.[18]

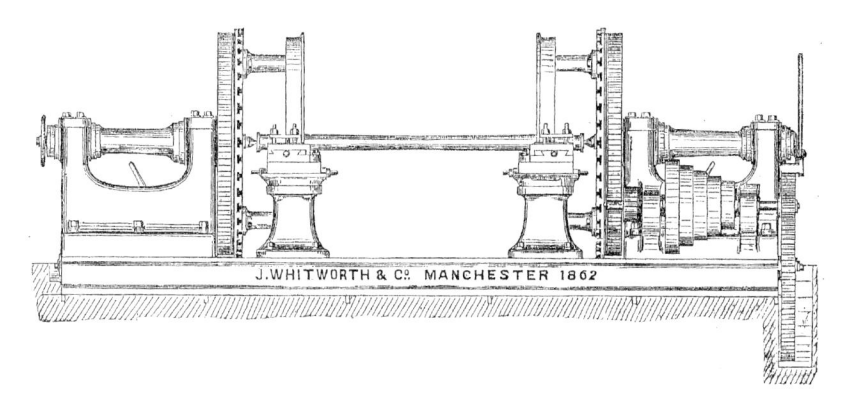

Fig.59 Whitworth's patent railway-wheel turning lathe. *(1862 Exhibition Catalogue)*

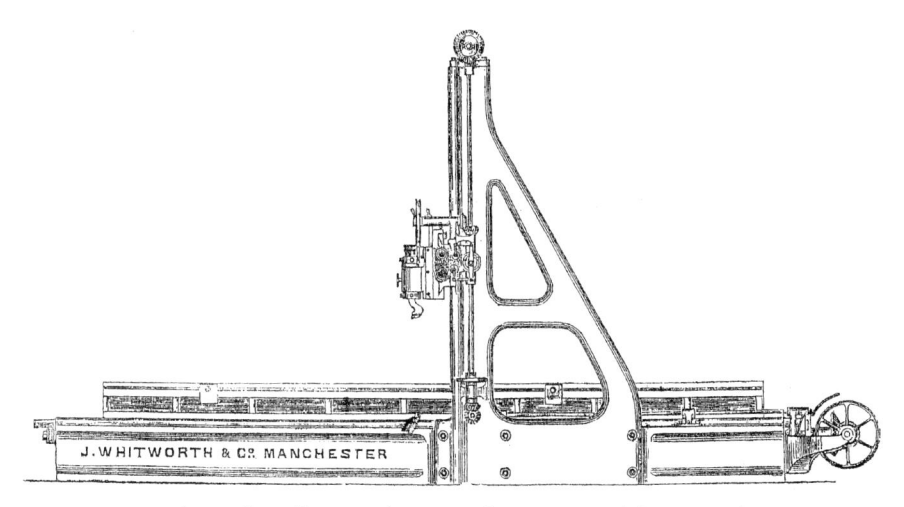

Fig.60 Whitworth's self-acting planing machine. *(1862 Exhibition Catalogue)*

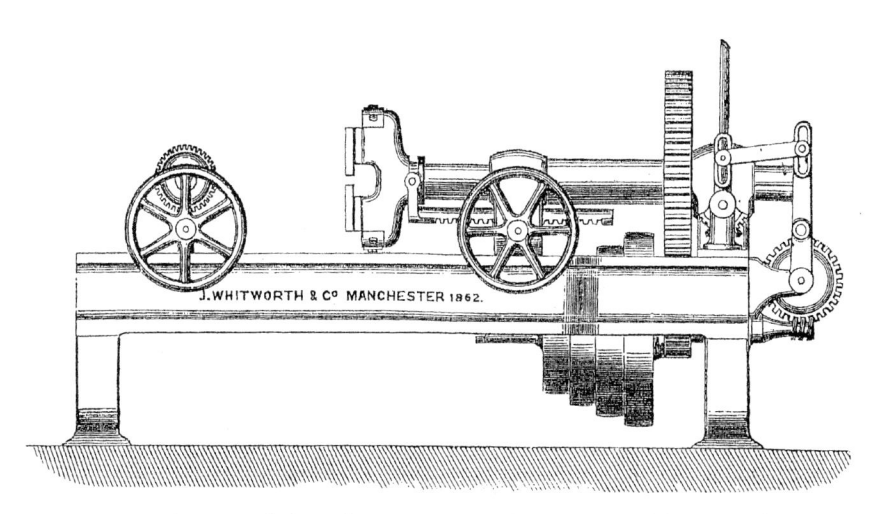

Fig.61 Whitworth's bolt- and nut-screwing machine. *(1862 Exhibition Catalogue)*

Fig.62 Whitworth's self-acting radial drilling and boring machine. *(1862 Exhibition Catalogue)*

Fig.63 Whitworth's self-acting compound slotting machine. *(1862 Exhibition Catalogue)*

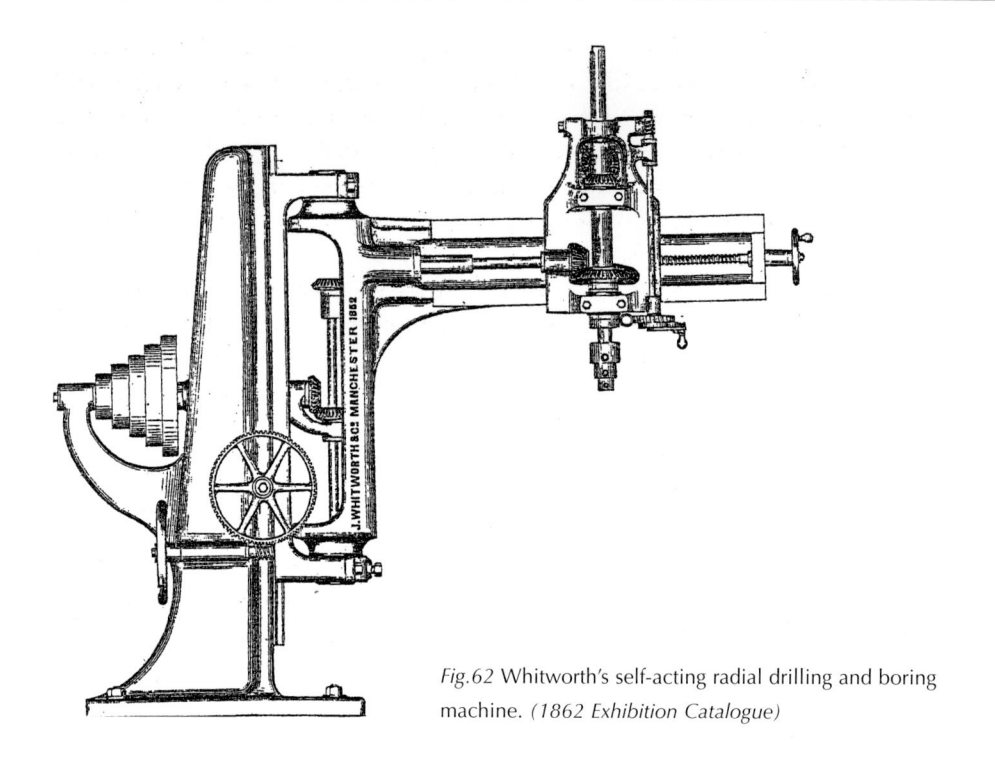

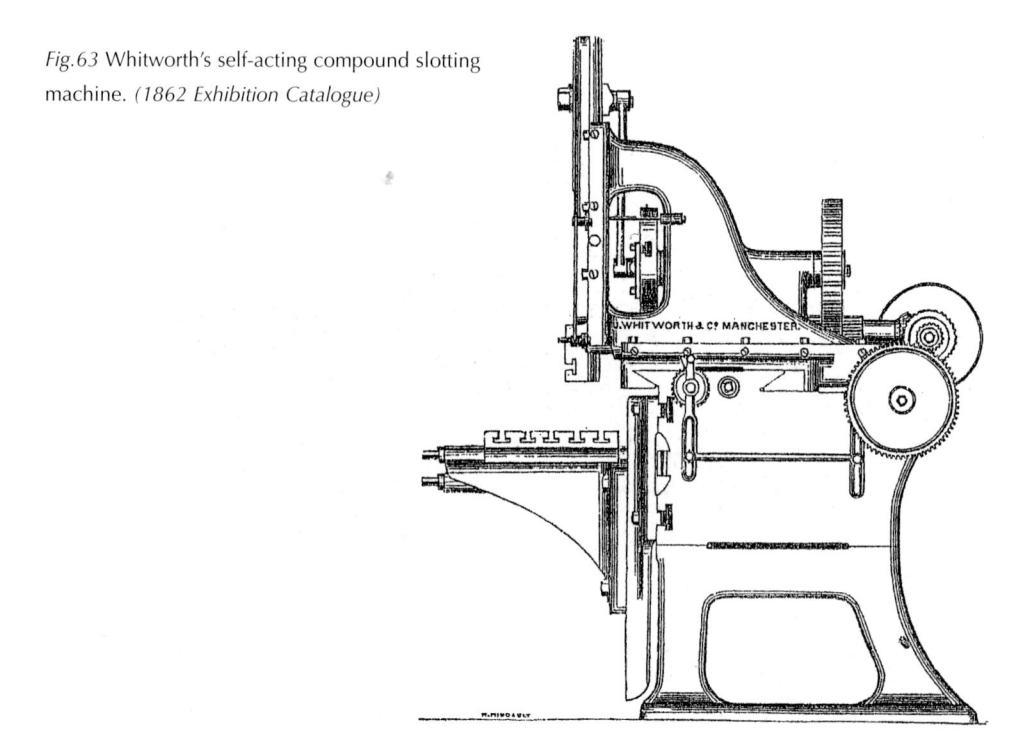

Whitworth was not as outspoken against the union as some other employers, such as Nasmyth and Fairbairn, but he went along with the closure until his workers had renounced the union.[19] Such was the animus against unionism at this time that the ASE was driven to defeat. Some of its members were assisted in emigrating to Australia, where they formed the first overseas branch of the union. Others made a formal renunciation while maintaining covert membership, or rejoined as soon as it became possible to do so without attracting unwelcome attention. One way or the other, the ASE survived, and the engineering employers came to be seen as having over-reacted to their own disadvantage. What perplexed Whitworth most was the apparent hostility towards mechanisation demonstrated by the men. With some employers, of whom Nasmyth was one, the goal of mechanisation was the elimination of the human workforce, so a certain defensive caution on the part of the unionists was at least understandable. Whitworth, on the other hand, perceived mechanisation as the source of an ever-expanding new field for skilled labour. He was impressed by the apparent fact that American workers so regarded it, and were in favour of further mechanisation rather than opposed to it on principle. He attributed this difference to a failure of understanding on the part of the British skilled workers, and thus to a lack of education. So he came to see education as a means towards enlightenment, whereby the workers would be enabled to recognise their genuine best interests and support him in his mission to expand the mechanical precision and standardisation of British industry.[20]

Whitworth's own record as an employer was not bad, but there was never any doubt in his establishments about who was in charge. He may have been a comparatively benign autocrat but, like most of his contemporaries in large British engineering businesses, he remained an autocrat. Eventually he used the limited liability legislation to encourage his workers to invest in their company by taking out shares. The scheme had some success, although only a few

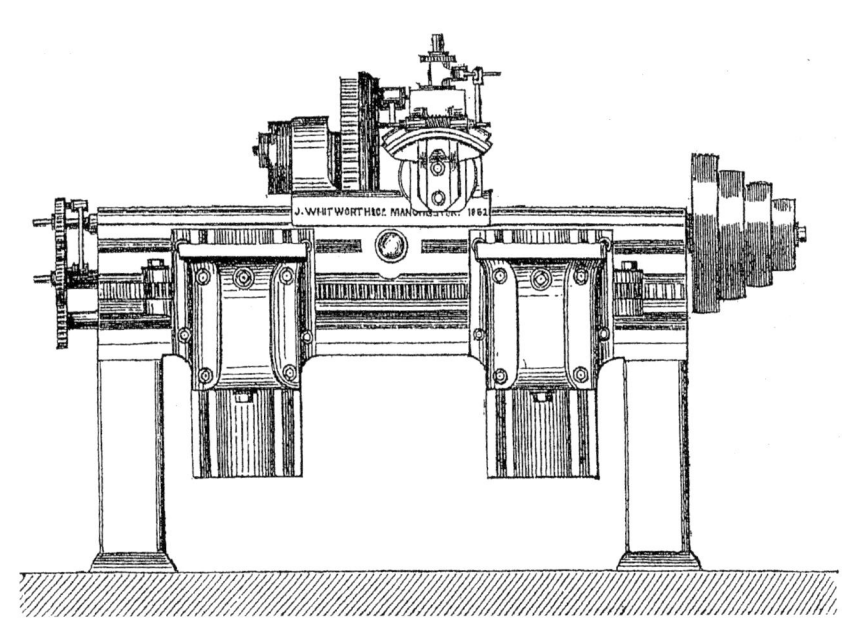

Fig.64 Whitworth's universal shaping machine. *(1862 Exhibition Catalogue)*

senior foremen could afford the initial investment, and Whitworth took care to retain the majority holding so that there was never any question of sharing managerial authority with the workers. Nevertheless, it was a tentative gesture of trust at a time when most British manufacturers demanded complete power 'to do what they will with their own'. The experiment did not appear to survive the merger, ten years after Whitworth's death, into Sir W.G. Armstrong Whitworth & Co. Ltd in 1897.[21]

With the British engineering industry back to normal, Whitworth could give some attention to the phenomenon of rapid industrial development in the USA, and to the educational lessons to be derived from it. He readily accepted an invitation from Lord Clarendon, Foreign Secretary in Lord Aberdeen's government, to join a Royal Commission to attend the American Exhibition of Industry, scheduled to be held in New York in May 1853. In the event, problems with the buildings delayed the opening of the Exhibition until well into July, but the commissioners went ahead with their visit in May. Their titular leader was Lord Ellesmere, but they appear to have been given separate briefs by the government and Whitworth's only close colleague was George Wallis, head of the Birmingham School of Art, and a well-known authority on small arms. There is little doubt that the government appointed Whitworth and Wallis in order to get authoritative information about the progress of the American armaments industry, and the two of them managed to arrange a full tour of American engineering and manufacturers of small arms in the north-eastern states. They stayed a few weeks and then returned before the Exhibition itself had opened. In his 'Special Report', written in conjunction with Wallis and published by the House of Commons on 6 February 1854, Whitworth set out the conclusions to his observations. In the first place, he was impressed by the size and variety of the American workshops, and their capacity to undertake many different types of engineering work. He found their machines heavy and robust, but not as yet up to British standards of precision, although he had no doubt that American industrialists and workers had both the will and the enterprise to overtake those of Europe in the near future. Ironically, Whitworth admired the comparatively open and democratic style of management practised in the USA, even though it was so different from his own practice. But, most significantly, he was struck by the willingness of the American labour force to co-operate in all mechanical innovations, and he attributed this quality to the superior education and free press available to everybody in the USA. He wrote:

> *The benefits which thus result from a liberal system of education and a cheap press to the working classes of the United States can hardly be over-estimated in a national point of view; but it is to the co-operation of both that they must undoubtedly be ascribed.*[22]

These observations made little immediate impact on the British government, but they were strongly felt by Whitworth and did much to determine one of the great themes of his career in promoting better standards of general and engineering education.

More immediately, however, Whitworth's trip to America determined the other main theme of the second part of his career – his commitment to armaments manufacture. The British government had become seriously concerned about the quality of armaments available to the army and navy, and this had been a primary reason for seeking the opinion of Whitworth, the outstanding British tool-maker, about the competition from American arms manufacturers. In October 1853, a few months after his American excursion and before he had submitted his report on that, the War Office approached Whitworth about the possibility of improving the Enfield rifle recently adopted by the Army. His first response was cautious. Like most informed observers, he was aware of the inadequacies of the small arms used by the military forces in Britain. The Enfield was an improvement on the 'Brown Bess', the musket that had done general service in the Napoleonic Wars, but it still used round shot which needed ramming down the barrel with every firing and had an effective range of only a few hundred yards. Whitworth appears to have been shocked by the mechanical inefficiency of these weapons and the method of loading and firing them, and immediately began to devise improvements. But he realised that, to make a thorough job of such improvements, it would be necessary to go back to first principles and work out the optimum conditions of performance for a totally new rifle. This would be an expensive operation in both time and financial investment, so before making any commitment Whitworth insisted on the necessity of being equipped with a full-scale enclosed firing range on which he could test his weapon. The War Office was so anxious to have his advice that it accepted his terms and built him the required range adjacent to his home in Fallowfield.

Once equipped with his firing range, Whitworth commenced in March 1855 on what Atkinson has called 'the two most fruitful years in the history of gunmaking'.[23] He worked systematically through a range of variations of barrel shape, size, bore, and rifling, and a similar range of projectiles, in an experimental programme that produced a far more efficient and reliable rifle than any known before. This was the 'Whitworth rifle', with its hexagonal bore of 0.45in and elongated hexagonal bullets designed to fit the barrel, deriving spin from the twist in the bore which gave greater stability in flight and hence precision in fire. Whitworth claimed no innovation in any part of this system, except for the machine tool – a twist drill with adjustable broach – that drilled the hexagonal bore. Westley Richards, one of the best gunmakers of the period, had made a polygonal barrel for I.K. Brunel in 1852, and Whitworth willingly deferred to this precedent. For his part, Brunel was happy that 'my friend Mr Whitworth' had developed the idea, although his family subsequently made a half-hearted claim for Brunel's priority in this invention.[24] But the complete system was Whitworth's creation, and in particular it was his success in working out by painstaking trial and error the best relationships between bore and bullet that placed the new rifle in a class of its own. When it was put to the test, in April 1857, the Whitworth rifle outclassed the Enfield in every respect. *The Times* reported that the test

> ...had established beyond all doubt the great and decided superiority of Mr. Whitworth's invention. The Enfield rifle, which was considered so much better than any other, has been completely beaten. In accuracy of fire, in penetration, and in range, its rival (the Whitworth) excels it to a degree which hardly leaves room for comparison.[25]

By the time Whitworth had demonstrated the superiority of his rifle, the Crimean War, which had provided an enormous stimulus to the drive for better weapons, was over. The competitive pressure to keep ahead of other European powers remained strong, but there had been changes of personnel at the War Office and a number of conflicting political initiatives stalled a decision about the Whitworth rifle. After much procrastination, a decision was taken in January 1859 not to proceed with the rifle. It is difficult to pierce the obscurity of the process by which this decision was reached, but it seems likely to have rested on the assumption that the Whitworth rifle would have been too expensive to produce in bulk, when there were cheaper, if less efficient, alternatives available for the British infantryman and tax-payer. The rifle, as a precision instrument, was undoubtedly expensive, but it was ideal for large-scale production because of its identical interchangeable parts, and this would have substantially reduced the costs of manufacture. But it was not to be, and the Whitworth rifle became a prized specialist weapon for sporting purposes and shooting competitions. The Queen opened the National Rifle Association meeting at Wimbledon on 2 July 1860 by firing a Whitworth rifle, pulling a silken cord attached to the rifle fixed in a frame and scoring a bull's eye. This demonstrated the consistent reliability of the rifle, but it did not convince the War Office that the weapon suited their needs.

Whitworth was understandably bitter at the treatment he received from the War Office over his rifle, but worse was to follow. He had quickly realised in the course of his programme of rifle tests that much of what he discovered about accurate rifling was applicable to larger weapons also. With the support of the War Office he began to experiment with artillery. He produced a most efficient 12lb cannon (the weight of the shell), and progressed to much larger weapons. Whitworth realised that the old style cast iron cannon was no longer appropriate for the more powerful weapons that he was now constructing. He turned to mild steel, only recently made available in bulk by the Bessemer process. He improved the strength of the steel by subjecting it to hydraulic pressure when still molten, thereby making a significant advance in metallurgical practice. By the time he had perfected his system for artillery, however, a formidable rival had appeared on the scene in the shape of William Armstrong, who was already manufacturing large guns for naval and military applications, based upon a system of coiling wrought iron around cast iron barrels. Armstrong, who had built up a large establishment at Elswick on the Tyne, had acquired a virtual monopoly of War Office provision by virtue of a network of personal contacts, which brought him a fortune, a knighthood, and subsequently a peerage. He had severe technical problems with his guns, and there was never any doubt that those of Whitworth were superior in quality. But they were also more expensive, and Armstrong was unscrupulous in exploiting his contacts in government, so that Whitworth was effectively excluded from this market also.[26]

These were bitter disappointments for Whitworth, but they did not prevent him from becoming a successful armaments manufacturer – he found significant overseas markets for his weapons and was able to produce them on a considerable scale. This development from a machine tool manufacturer to a manufacturer of armaments does not seem to have troubled Whitworth unduly. He was able to delude himself with the specious argument, used of other technical advances before and since, that having made weapons that were so effective, nobody would dare to use them. But it did lead to some difficulties with pacifists such as John Bright, with whom he had previously had a close and friendly relationship, although there does not appear to have been an open quarrel between the two men. Perhaps of more general significance to Whitworth's career, his armaments added to his fortune which he had already made with his machine tools. The disposition of this fortune was a subject of great concern to Whitworth in his closing decades, and provided the basis of his lasting reputation as a philanthropist and educational reformer.

Whereas Armstrong achieved the elevation of his social status as a result of his success in the arms trade, his rival Whitworth received his baronetcy for his services to education. As we have observed, Whitworth had already established himself in the social and intellectual life of Manchester, and had identified himself with many Mancunian movements of social reform. These culminated in the creation of Owens College in 1851 and its development as a science-based university-style college in 1867, with Charles Beyer and Whitworth leading the enterprise. 'Owens will be a science-orientated modern liberal rival to the ancient universities' they said, and secured the appointment of Osborne Reynolds to be the first professor of engineering at the College.[27] The College flourished, going on to become the Victoria University and eventually the University of Manchester. Whitworth made further generous gifts to the institution, including his house in Fallowfield, The Firs, which was first made available to C.P. Scott as his home while he was editor of the *Manchester Guardian*, and then incorporated into the University as the Vice Chancellor's Lodge. The lifetime loan to Scott reflected another of Whitworth's long-standing interests: the value of good newspapers in raising the general standard of education in the community.

But Whitworth made an even more specific benefaction to the cause of education in the shape of the 'Whitworth Scholarships', launched in 1868. On 18 March that year Whitworth wrote to the Prime Minister, Disraeli:

> *I desire to promote the engineering and mechanical industry of this country by founding 30 scholarships of the annual value of £100 each, to be applied for the further instruction of young men, natives of the United Kingdom, selected by open competition for their intellectual proficiency in the theory and practice of mechanics and its cognate sciences... I venture to make this communication to you in the hope that means may be found for bringing science and industry into closer relation with each other than at present obtains in this country...*[28]

The offer was gratefully accepted and the first scholarships were awarded in 1869, with ten candidates all receiving £100 a year for three years. Whitworth suggested that the fund should be administered by the government, and the responsibility devolved upon the Department of Education after the Education Act of 1870. Then, under the terms of his will, a capital sum of £100,000 was given to endow the foundation after he died, and grants of scholarships and prizes have been made from this fund regularly ever since. By 1926, over 1400 men had benefited directly from Whitworth's benefaction, and the number has continued to grow steadily.[29]

Whitworth had hoped to consolidate his contribution to technical education in Britain by leaving his house in Darley Dale as a college of engineering, but this plan was not fulfilled. He was prevented from donating the house for this purpose because at his death in 1887 the government did not possess the powers under the Education Act to promote technical education. His widow continued to live in Stancliffe Hall, and it eventually became a school. Nevertheless, his contribution to the campaign for providing good quality technical education in Britain was enormous. He was among the first spokesmen in the country to make it a matter of public concern, and he consistently urged its necessity among colleagues and on public occasions. He allied himself on this matter with campaigners such as Henry Cole, Lord Playfair and John Scott Russell. After the Paris Exhibition of 1867 he joined the chorus of those who blamed the comparative failure of British products to make a good impression on the educational deficiencies of British workmen and managers.[30] His stand on this subject, and his readiness to put large sums of money into solving the problem, was greatly appreciated by public opinion, leading to the award of a baronetcy in 1869. The hundreds of young men who benefited substantially from his generosity became an elite of British technical education, proud to call themselves 'Whitworth Scholars', and many of them went into a new profession teaching science and technology.

Sir Joseph Whitworth died at Monte Carlo on 22 January 1887. The distinct change of direction which had characterised his professional activities after the Great Exhibition in 1851 had been mirrored in his private life by an increasing distance between himself and his first wife, Frances, and by his purchase of the Stancliffe estate in Derbyshire. He had married Frances Ankers, the daughter of a Cheshire farmer, on his way to London in 1825, and she had accompanied him throughout his years of training in the capital, and then the years of grindingly hard work while he built up his machine tool enterprise in Manchester. They had eventually established some physical comfort for themselves in The Firs at Fallowfield, but they had no family, and in the 1850s Frances seemed to drop out of the social invitations that came to Whitworth. She appears to have had no part in his decision to acquire Stancliffe Hall in Darley Dale, which he purchased in 1856 for the sum of £33,850.[31] This house had a

Fig.65 Whitworth letter and attached portrait. *(Reproduced by courtesy of the Director and Librarian, The John Rylands University Library of Manchester.)*

somewhat unprepossessing situation in the middle of a heavily quarried area, but Whitworth devoted his customary meticulous attention to the improvement of the house and the development of the estate. He made his regular home in Darley Dale, leaving Frances at The Firs and presumably seeing her only in the course of his regular business trips to Manchester. During these years he acquired an attachment for Mary Louisa, daughter of Daniel Broadhurst, Town Treasurer of Manchester, and widow of Alfred Orrell, who had been Mayor of Stockport. Within six months of the death of Frances, in 1870 – by which time she appears to have gone to live with her family at Delamere in Cheshire – Sir Joseph had married Mary Louisa, who thus became the second Lady Whitworth at the age of forty-three, while her new husband was sixty-eight.

The Whitworths applied themselves vigorously to the role of leading citizens of Darley Dale, building houses on the estate and making donations to local schools and other charities. Whitworth quarrelled with the vicar of St Helen's parish church, but this did not prevent him from being buried in the churchyard there, to be joined in 1896 by Mary Louisa. He left handsome sums of money for the construction of the Whitworth Institute in Darley Dale, opened in 1890 as a centre for education and community activities, together with a cottage hospital and various other good causes in the village. Whitworth avoided many problems in the apportionment of his fortune by leaving it all to his wife and two trusted friends, with detailed instructions about how he would like it to be spent. The three executors carried out his wishes to the full, with benefactions to educational causes in Manchester and the creation of the Whitworth Art Gallery as well as the gifts to Darley Dale. He was one of the first great industrial philanthropists, who succeeded in modifying the landscape and contributing to the transformation of society by the generous bestowal of the fortune he had made in manufacturing industry.[32]

Whitworth's reputation, however, was only incidentally that of a philanthropist. First and foremost, he was a successful businessman, an innovator of genius who established the highest standards of precision in machine tool making, and persuaded the whole of the British engineering industry to adopt the standards of uniform screw threads, workshop precision, and decimalization which he had made his own. His machine tools dominated British mechanical engineering practice from the 1840s until the end of the century.[33] In addition to this colossal achievement, he made another reputation for himself as an armaments manufacturer of outstanding quality, producing rifles and artillery of hitherto unimaginable precision and reliability, and solving with brilliance the metallurgical problems which he encountered in the process. He brought the standards of scientific instrument making to an industry that had previously relied more on the craft of the blacksmith. The War Office – and the British soldier – was prevented from benefiting immediately from these superior weapons partly by their cost, but mainly because Whitworth was outmanoeuvred by his rival Armstrong, who possessed a more effective network of supporters. But within two decades, the standards which Whitworth had built into his weapons were generally accepted, even if their acceptance completely failed to fulfil his hope that they would not be used in anger.

Whitworth then went on to make another reputation for himself as an educationalist of vision. Although he was not alone in recognising the inadequacies of the British educational system – particularly the blatant lack of provision for even the most basic education for the mass of the population, and of any sort of instruction in the scientific principles of practical engineering and industry – unlike most of his fellows he was prepared to do something about it. He devoted a tremendous amount of time and wealth to improving technical instruction through mechanics institutes, the university-style institution created in Owens College, and the Whitworth Scholarship scheme. He seems to have instinctively felt that Britain, having led the world in industrialisation in the first half of the nineteenth century, and having become in effect the 'workshop of the world', was in danger of forfeiting this pre-eminence through the ignorance of its workers and managers. Partly as a result of his own achievements, he saw more clearly than most of his contemporaries that it was urgently necessary for Great Britain to overcome its prejudices about state intervention in a laissez-faire regime when standards of precision and scientific accuracy had become so important to industrial progress. Britain needed now to equip itself with a sound educational system that would enable its citizens to compete with those of other countries committed to rapid industrialisation. In the short run, it failed to do so and paid the penalty by sacrificing industrial leadership to other countries. But in the long run, Whitworth's great contributions to educational improvement bore fruit, and it was because of him, and others like him, that Britain remained in the competition.[34]

Whitworth made fortunes both as a machine tool manufacturer and as an armaments manufacturer. Like most successful Victorian engineers and entrepreneurs, he used part of his hard-won wealth to establish himself in comfortable houses with fine gardens. But he retained great personal frugality, so that plenty remained to distribute in his will, and hence his reputation as a philanthropist. However well deserved, this reputation is only the icing on the cake. Whitworth deserves to be remembered as the man who placed the machine tool industry on a scientific basis, thereby preparing the way for mass production, automation, and the huge achievements of the modern engineering industry. In this, as in other respects, his contribution was pivotal. By providing the best machines possible in the old systems of production that had served Britain so well in the first half of the nineteenth century, he supplied the means for the transfer to a new system based on mass production and interchangeability of parts. The likelihood that this new system would be pioneered in America rather than in Britain was recognised by Whitworth, and this made him conscious of the educational deficiencies of the British worker in contrast with his American counterpart. In fact, American practice in machine tool operation rapidly surpassed that of Whitworth after the end of the Civil War, so that some later American commentators such as Charles Porter were able to write disparagingly about Whitworth's performance, even while acknowledging Whitworth himself as 'a phenomenal man'.[35] But without Whitworth's machines, the adoption of the new production techniques in America which provided the basis of the modern engineering industry would have taken much longer. Above all, it was the achievement of Whitworth to determine the huge impact that engineering has made in modern industry and in the modern world.

Endnotes

[1] There is a dearth of primary material about the life of Whitworth. It seems that Lady Whitworth destroyed most personal and business papers on the death of her husband. The small collection of papers by B.R. Faunthorpe in JRUL contains a few documents relating to the Stancliffe estate, but is mainly secondary material. Atkinson 1997 is a useful compendium of information. See also Musson

1. 1963, 1966 and 1975; *DNB*; Lea 1946; Kilburn 1987; and obituaries in *PICE* 91 (1887-88) pt.1, pp.429-46; *PIME* (February 1887) pp.152-56.
2. Carlyle (ed.) 1903, Letter 78.
3. I have followed Atkinson 1997, Chapter 1, 'The Making of a Baconian Engineer' in this summary. The traditional account describes Whitworth as the son of a schoolmaster, and as being sent to school in Bradford. But it is more probable that his father was a craftsman in the textile industry, and that it was he who was sent to the Bradford school as part of his preparation for his ministry in the church.
4. The story is told in several forms, but see the obituary of Whitworth in *PICE* (1887-88).
5. Smiles (ed.) 1883, Chapter 8, 'Maudslay's Private Assistant'. Smiles did not give much biographical attention to Whitworth, but there is a brief note in Smiles 1863, pp.273-74, of one paragraph starting: 'Mr Whitworth is another of the first-class tool-makers of Manchester who has turned to excellent account his training in the workshops of Maudslay and Clement'.
6. Patent 8475, 1840. William Muir was employed by Whitworth to do the drawings for this invention, and it is possible that he deserves some of the credit for it. Whitworth contributed a substantial paper on 'Mechanical Street Cleaning' to *PICE*, 6 (1847), pp.431-65; unusually, the article was accompanied by some graphic drawings of street scenes, and Edwin Chadwick made a long contribution to the discussion.
7. Engels 1958. Engels subsequently demonstrated an interest in the Whitworth rifle: see his seven-part 'History of the Rifle' in the *Lancs & Cheshire Volunteers' Journal* (1860-61). I am grateful to Andy Buchanan for this reference.
8. For the Holtzapffels, see Rolt 1965, pp.35, 88, 100, etc.
9. In 1834 Whitworth's wages bill at Chorlton Street was almost £50 a week, accounting for about twenty workmen (Atkinson 1997, p.49). By 1854, the work force had risen to 368, and by 1874 to about 750. Whitworth's larger factory at Openshaw, established in 1880, employed over 1000 by 1884: see Kilburn 1987, p.33.
10. *PICE*, 1 (1841), pp.157-60: the patronising tone of the comments reported in the discussion by Field, Seaward and the President (Walker) suggest that they thought that Maudslay had done it all before, although they expressed gratitude to Whitworth for his presentation.
11. *PIME* (1856) p.125; (1857), p.134.
12. Sir Joseph Whitworth, *Papers on Mechanical Subjects* (London and Manchester, 1882), reproduces these papers in Part 1, 'True Planes, Screw Threads and Standard Measures'. An earlier collection of some of these papers was published as Joseph Whitworth, *Miscellaneous Papers on Mechanical Subjects*, (London, 1858).
13. Kargon 1977.
14. Pole (ed.) 1877, pp.263-69.
15. Timbs 1851.
16. JRUL, autograph letter to Mr Rawlinson, 23 July 1873; a faded photograph of Whitworth is stuck to this letter (see Fig.65). I am grateful to Dr Brenda Buchanan for locating this specimen.
17. Atkinson 1997, Chapter 1.
18. S. and B. Webb 1894, Chapter 4, 'The New Spirit and the New Model'; see also Buchanan 1957.
19. Nasmyth, quoted by Cantrell 1984, p.224, attributed his early retirement to his impatience with trade unionism. Fairbairn's son Thomas wrote a series of anti-union letters to *The Times* as 'Amicus'.
20. The background to this recognition of differences between British and American practices has been illuminated by Habakkuk 1962; see also Saul (ed.) 1970.

[21] Atkinson 1997, pp.298-99, on industrial relations and the merger with Armstrong.

[22] House of Commons, Report of Joseph Whitworth, 6 February 1854.

[23] Atkinson 1997, p.213.

[24] I.K. Brunel had considered the possibility of polygonal barrels as early as 1852, as he wrote to Westley Richards, the Birmingham small-arms manufacturer, on 7 February 1853 requesting: 'I want a rifle barrel made octagon shaped inside…' and there are subsequent letters on 9 December 1854, 21 July 1855, 22 May 1858, and 26 November 1858, in the last of which he tells Richards: 'I have no intention of interfering with Mr Whitworth's patent…' (see the *Private Letter Books* of I.K. Brunel, Bristol University Library). This disclaimer did not prevent Brunel's son from making a claim of priority in this invention on his father's behalf: see Brunel 1870, pp.449-52.

[25] *Times*, 23 April 1857.

[26] Tennent 1864 and 1972 gives a detailed presentation of Whitworth's case.

[27] Reynolds had a long and distinguished career at Owens: see McDowell and Jackson (eds.) 1970; also Buchanan 1989, p.171.

[28] Low (ed.) 1926 lists all scholars to this date, and gives an introductory memoir and relevant documents: for the letter to Disraeli, p.27.

[29] Low (ed.) 1926, p.v. A subsequent edition appeared as *The Whitworth Register 1998*, edited for the Whitworth Society by F.M. Burrows. The pattern of prizes has been complicated by increased provision of state support for higher education, but the range of awards under the Whitworth scheme has continued to grow and to be widely appreciated. I am grateful to Professor Burrows and G.M. Ward for their help in up-dating me with this information.

[30] Ashby 1959 gives an account of the impression made by the Paris Exhibition of 1867 and quotes the report by Dr Lyon Playfair in an Appendix, pp.111-113.

[31] Kilburn 1987, p.39.

[32] Kilburn 1987, Chapters 5 and 6; Atkinson 1997, Chapter 10. There is a file in JRUL, Faunthorpe collection, on the law suit in which distant relatives of Whitworth challenged the allocation of his fortune, ending with the judgment in 1921, reported in the *Times,* 7 April 1921, which dismissed the action.

[33] An interesting sequence of letters in BCA, Boulton & Watt papers, follows the construction and installation of a lathe made by Whitworths in 1842: Box 5/II/151-161. I am grateful to Mr T.J. Procter for this reference.

[34] The thesis that British industry was in some sort of decline in the second half of the nineteenth century has been widely adopted and expressed in most dramatic form by Wiener 1981. But economic historians have generally taken the view that the 'decline' was relative, as Britain was overtaken by other manufacturing countries. For this view, see Landes 1969; also Rubinstein 1994 for a more specific critique of the Wiener thesis.

[35] Porter (1985) also described Whitworth as a 'monumental egoist' and was sharply critical of some of his workshop practices.

8
James Nasmyth
John Cantrell

When James Nasmyth entered the employment of Henry Maudslay in May 1829 at the age of twenty, he was anything but an engineering novice.[1] He had already constructed a road steam carriage capable of carrying eight passengers which made successful trials on the Queensferry Road near Edinburgh, together with a number of model steam engines for private sale.[2] He had also published an account of his expansometer, an instrument for measuring the total and comparative expansion of solids, in the *Edinburgh Journal of Science*.[3] He had become an associate member of the Edinburgh Society of Arts in December 1826[4] and had made a succession of communications to the Society including descriptions of a high pressure steam engine with a method of applying waste steam to increase the temperature of the furnace.[5] In 1829 Nasmyth was awarded the Society's silver medal for his method of easing the motion of complex pulleys.[6] Indeed Nasmyth's admission to Maudslay's employment was secured only after submitting specimens of his work including a working model of a high pressure steam engine and various examples of mechanical drawing and draughtsmanship. Nasmyth claimed that he needed to convince Maudslay that he 'was not an amateur, but a regular working engineer'.[7]

Fig.66 James Nasmyth by D.O. Hill, c.1845. *(Scottish National Photography Collection)*

Hence Nasmyth did not come to Maudslays to learn a trade but rather to consolidate his engineering practice and gain insight into the methods of managing a large-scale engineering works. Accordingly, he was admitted not as an apprentice but as Maudslay's 'assistant workman' to be based in his master's private workshop. This favoured position reflected the relative privilege of Nasmyth's background and upbringing. Born in 1808, James Nasmyth was the tenth child and fourth son of Alexander Nasmyth (1758-1840), the celebrated Scottish portrait and landscape painter.[8] His mother, Barbara Foulis (1765-1848), though without fortune, was the sister of the baronet Sir James Foulis. The Nasmyth household, based at 47 York Place, Edinburgh, was a lively, stimulating and comfortable home for the young James. It supported a nursemaid and servants and was frequently the meeting place for some of the most eminent scientists, artists and literary men in Edinburgh. Sir Walter Scott, James Watt and Sir Henry Raeburn were among Alexander Nasmyth's friends and visitors.

Nasmyth's early formal education was both unhappy and unsuccessful and he left his first school in George Street in 1817 following a serious physical assault by his teacher. In 1821 or 1822 he left Edinburgh High School where he 'made but little progress' knowing 'a small amount of Latin and no Greek'. A combination of ill temper from his tutor, learning by 'rote and cram', and classes of nearly two hundred boys was more than enough to extinguish Nasmyth's interest in classical learning.[9] The valuable part of his early education was achieved

THE ROAD STEAM-CARRIAGE. BY JAMES NASMYTH.

Fig.67 Nasmyth's road steam carriage. *(Nasmyth, 1883)*

at home and in the company of his friends. As a result of Alexander Nasmyth's proficiency in both architecture and mechanics he was able to offer his son lessons in draughtsmanship and the opportunity to learn the rudiments of practical mechanics in his York Place workshop. The latter was equipped with a small furnace stove, an anvil, tongs, foot lathe and bench. Nasmyth called it his 'primary technical school' and the 'very foreground' of his life. He was soon sufficiently skilled to begin producing spinning tops and small brass cannon, and by the age of seventeen was able to construct his first steam engine for mixing oil colours. Further useful practical experience was obtained in an iron foundry and chemical laboratory owned by the fathers of two school friends: at Patterson's foundry, Nasmyth was able to see how iron castings were made and learnt the art of hardening and tempering steel by plunging the red hot metal into cold water; and in Tom Smith's laboratory Nasmyth experimented in the production of acids, alcohol, sulphuric ether and phosphorus.[10]

Theoretical instruction was not abandoned following Nasmyth's departure from the Edinburgh High School but became more directly linked to his interests and aptitudes. In 1821 he enrolled at the Edinburgh School of Arts, the forerunner of Heriot-Watt University, and attended evening classes in chemistry, mechanical philosophy, geometry and mathematics. This was supplemented by daytime studies at private classes where he was introduced to Euclid's *Elements*. With part of the proceeds of the sales of his model steam engines, sold for £10 each, Nasmyth purchased tickets to a number of University lectures, attending courses in chemistry, geometry and mathematics and natural philosophy.[11]

At home his engineering activities became increasingly ambitious and included at least two full size engines: a direct-acting high pressure engine with 4in diameter cylinder for driving the large turning lathe of a local iron worker, George Douglass; and a 2hp engine for driving the machinery of a braiding manufacturer. As a small scale steam engine maker himself Nasmyth was keenly interested in the engines employed by local breweries, distilleries and other establishments. He inspected these engines and compared the workmanship of the different steam engine manufacturers. He soon became impressed with the superiority of those made by Carmichaels of Dundee and upon further enquiry established that the firm made use of improved machine tools and employed a number of leading men who had previously worked at Maudslay's Lambeth factory.[12] So began Nasmyth's admiration for Henry Maudslay and an overriding ambition to broaden his knowledge of machine tool technology at the hands of one of the acknowledged masters of the craft.

By the time that Nasmyth set sail for London in May 1829 he was not just a precociously talented engineer but a young man with broad and developing scientific interests including chemistry, geology and astronomy.[13] This was largely as a result of the influence of his father and his father's friends. Nasmyth recalls being party to scientific discussions between David Brewster, James Hall and John Leslie concerning the volcanic origin of the Edinburgh scenery and other matters.[14] With his father's acromatic spy-glass of 2in diameter Nasmyth was able to gaze at the moon, planets and sun spots. Inspired by his father's accounts of what he had seen through a 12in reflecting telescope, he resolved to make his own instrument and in the course of constructing an 8in reflecting telescope on the Newtonian plan, devised a new method of casting specula so as to ensure freedom from defects and an enhanced brilliancy.[15] Nasmyth's wide ranging scientific and artistic interests were to have a profound influence on the course of his career. As early as 1845 he was looking forward to the time that he could down tools and indulge himself in what he termed 'hobby-riding'.[16] Engineering was to Nasmyth a means of securing the resources to pursue his wider interests in art, nature and the heavens.

It is of little surprise that Henry Maudslay was impressed by someone who possessed such practical skill and intellectual curiosity. There were perhaps more personal reasons why Nasmyth was engaged by the great engineer: he admits that his father had been introduced to Maudslay by a mutual friend in 1826 or 1827. Alexander himself refers to the lately departed Maudslay as 'my worthy friend' in a letter of 1832.[17] Since Alexander Nasmyth was often in London and was the inventor of both the 'bow and string bridge'[18] and a method of riveting by compression, areas likely to interest Maudslay, there would have been some foundation for a friendship to blossom. Certainly the relationship between James Nasmyth and Maudslay soon became a close one, with Maudslay apparently entrusting Nasmyth with his life story.[19]

Nasmyth began work at Maudslays on 30 May 1829 and remained there, beyond Henry Maudslay's death in February 1831, until August 1831. During this period of some twenty-six months Nasmyth was involved in a variety of work. His first tasks were to assist Maudslay in making some modifications in the latter's screw-originating machine, followed by the construction of a model of a pair of 200hp marine engines. The latter required some 300 minute bolts and nuts. Nasmyth devised what became known as his nut-cutting machine to produce the hexagonal nuts, an early example of the milling machine. He was also engaged in the construction of large models of ships' hulls for testing their resistance to passage through water when propelled at different velocities. The recording of the practical correctness of the different designs was achieved by specially contrived instruments prepared by Nasmyth. Other tasks included working alongside James Sherrif in overhauling and improving a French machine for use at the Royal Mint; making a perspective drawing of some 200hp engines to be purchased by the Admiralty for the *Dee*; and casting an 8in diameter speculum for Maudslay's telescope. One novel contrivance devised by Nasmyth in 1829 was a mode of transmitting rotary motion by means of a flexible shaft formed of a coiled spiral wire; this was to enable holes to be drilled in inaccessible places where use of the ordinary drill was impossible. A similar arrangement was later adapted for dentists' work. Nasmyth's final assignment for the Lambeth factory, performed for Joshua Field, involved assisting him in making working drawings of a 200hp condensing steam engine for the Lambeth Waterworks Co.

The majority of Nasmyth's work at Maudslay's appears to have been that of draughtsman and maker of models and instruments. Such small scale work was very similar to his Edinburgh activities but in Lambeth he was also able to observe the machine tools in the machine shop, discuss the latest improvements with Maudslay, taking notes and making sketches where necessary, and watch his methods of production management and control. Nasmyth wrote that 'the entire establishment thus became to me a school of practical engineering of the most instructive kind'. He was also able to extend his acquaintance with the wider scientific and engineering community, meeting Michael Faraday, Samuel Bentham, Bryan Donkin and John Barton of the Royal Mint.

In September 1830 Nasmyth took a three week holiday from Maudslay's, partly to attend the opening of the Manchester to Liverpool Railway. It was during this excursion that he first visited Patricroft and determined to base his business activities in the north west of England.[20]

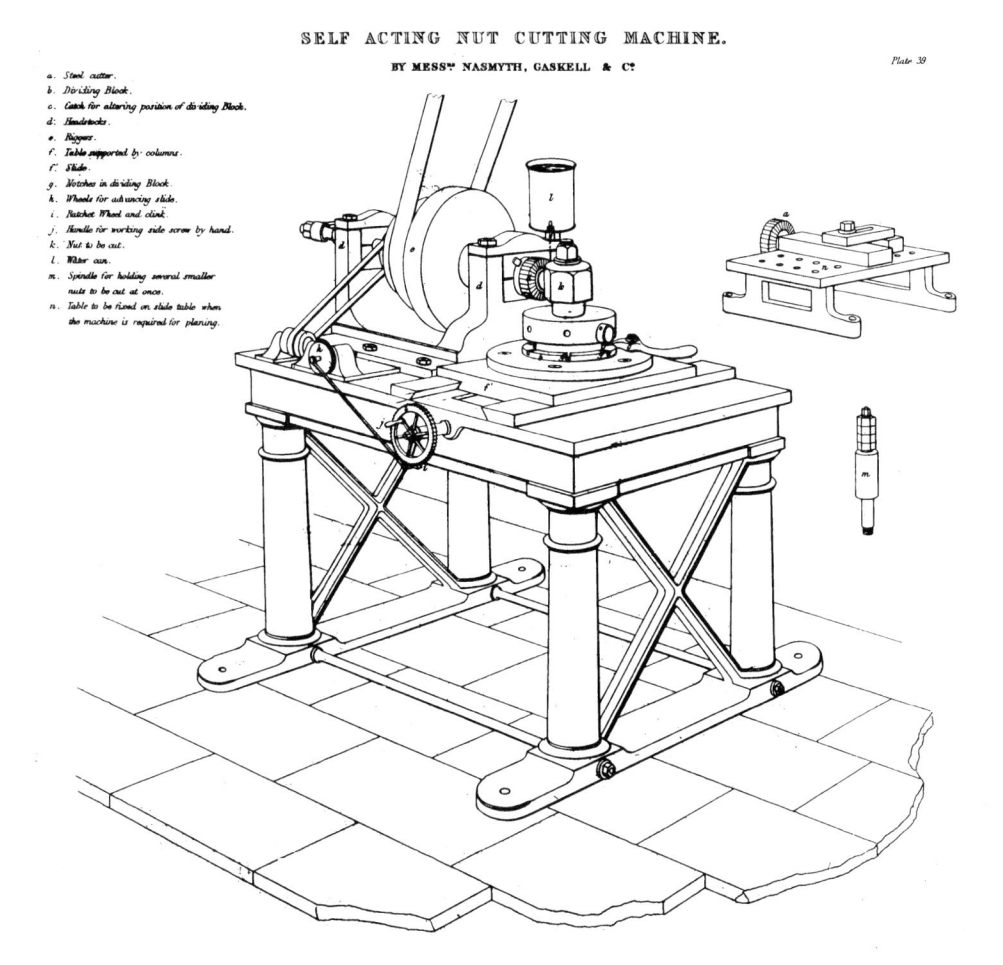

Fig.68 Nasmyth's nut-cutting machine. (Buchanan, 1841)

Nasmyth did not, however, move directly from Maudslay's to Manchester in August 1831. It was approximately three years before he set up business in Dale Street. In the meantime he was primarily occupied with the manufacture of tools to enable him to begin independent business. This activity was carried out in a temporarily constructed workshop at Old Broughton, Edinburgh, some five minutes walk away from York Place. Before leaving London, Nasmyth was allowed to obtain castings of one of the best turning lathes in the Lambeth workshops. This became, in Nasmyth's words, 'the parent of a vast progeny of descendants ... in planing machines, screw cutting lathes, and many other minor tools'.[21] An important part of Nasmyth's debt to Maudslay was therefore of a tangible kind. The technology employed by Nasmyth during his early years in business was quite literally descended from that of his former master.

One of the salient features of Nasmyth's business career is its brevity. He retired in December 1856 at the age of forty-eight having made himself, in modern terms, a multi-millionaire.[22] Yet he began twenty-two years before with just £63 capital. The secret of this success can be largely attributed to two extraordinary factors: Nasmyth's ability to attract substantial investment capital at an early stage in his operations;[23] and the invention or promotion of the steam hammer which enabled Nasmyth to reap super-normal profits for fourteen years as a result of his 1842 patent.[24] The two factors are linked in that the invention of the steam hammer was a response to the demand for the heaviest class of engineering work which depended upon a large-scale manufacturing base. The Bridgewater Foundry, Nasmyth's handsome custom-built factory, was largely financed by outside capital.

Initially, Nasmyth's engineering operations were quite modest. In about June 1834 he set up shop in a factory flat occupying the first floor of a disused cotton mill measuring 130ft by 27ft in Dale street, off Piccadilly, Manchester. The landlords, Wren & Bennett, charged an annual rent of £50. At this stage there were no sleeping partners though Nasmyth had offers of credit facilities amounting to a total of £1500 from the calico printer and cotton manufacturer William Grant and the banker Edward Loyd. Orders poured in and Nasmyth was soon busy producing small steam engines, parts for printing machines, and flat surfaces with his planing machine. A smithy established in the basement flat of the mill together with the services of a nearby foundry enabled him to engage in a wide variety of engineering work. But the premises were crammed. In August 1835 Nasmyth wrote to his friend D.O. Hill, 'we are as full of employment as we can be from the disadvantageous circumstances under which we are working as regards premises of sufficient accommodation and extent'.[25] In the same letter Nasmyth reveals that he expected to be moving 'very shortly to some new situation' and that it was 'necessary to conect [sic] ourselves with some one of capital in a partnership'. The departure from Dale Street to other premises became an urgent matter when the beam of a 20hp engine crashed through the floor of the factory flat on to the premises of a glass cutter located below.

Fig.69 Nasmyth's Dale Street workshop. *(Nasmyth, 1883)*

For the rest of his business career Nasmyth was based at Patricroft where the Bridgewater Foundry was built close to the intersection of the Liverpool to Manchester Railway and the Bridgewater Canal. It appears that some extra source of finance became available in July 1836 and by October, Holbrook Gaskell (1813-1909) had entered the partnership 'possessed of a moderate amount of capital' which amounted to £4610 by the end of June 1838.[26] Gaskell took control of the commercial operations of the firm. The main source of finance was soon to be the Birley family, influential cotton manufacturers with broad business interests, who injected more than £40,000 into the Patricroft venture by the middle of 1838. When they withdrew in the same year their place was taken by Edward Loyd's solicitor, George Humphrys, and Henry Garnett (1814-1897), the third son of Robert Garnett, a prominent Manchester cotton merchant and railway director. This investment enabled Nasmyth to design and construct one of the largest engineering works in the country: smithy, ball furnace, brass foundry, engine and boiler house, grinding room, main foundry, planing and heavy turning shops, locomotive machine-shop, erecting shed and pattern store. There were also a number of dwelling houses especially erected to house employees. Items of equipment within the works in 1837 included two foundry cranes, a foundry weighing beam and scales, a carriage and railway for the cupola and a wharf crane. All this enabled Nasmyth to begin competing with the most well established engineering firms within seven years of leaving Maudslay's.

Before the advent of steam hammer production in 1843 Nasmyth concentrated on the manufacture of machine tools, steam engines and locomotives. A printed price list was issued in 1839 advertising multiple sizes of planing machine, drilling machine, turning lathe, foot lathe and slide lathe. There were also nut-cutting, shaping, slotting and key-grooving, wheel-cutting, punching and shearing and plate bending machines. Some of these tools were of

Fig.70 The Bridgewater Foundry. *(Nasmyth, 1883)*

enormous proportions: the great key grooving and paring engine had a 3ft stroke and was capable of accommodating wheels of 12ft diameter, whereas the great vertical boring mill had a 30in diameter boring bar and could bore any cylinder under 14ft diameter.[27] All these machines were designed by Nasmyth Gaskell & Co. and Nasmyth stated that 'each new tool that I constructed had some feature of novelty about it'.[28] Sometimes this was achieved by making 'a carefull inspection and selection of the best parts of many such machines in use hereabouts'.[29]

Among the tools in the printed list was one patented invention, the patent slotting and key-grooving machine. The purpose of this machine was to cut key seats in the eyes of wheels or pulleys to accommodate the wedge-shaped keys that linked shafts to their attachments such as flywheels, gears, or pulleys, so preventing relative rotary motion. According to Nasmyth it was Richard Roberts who first mechanised this process, previously performed by hand with hammer, chisel and file, by adapting Maudslay's mortising machine for block-making into a machine for operating upon metal.[30] In Roberts' machine the positioning of the tool, bearings and driving apparatus was above the table and this 'limited the extent of work or the diameter of the wheels to be operated upon...'. In Nasmyth's machine the table was placed 'above the whole of the other necessary parts and driving apparatus ... and the cutting tool is caused to rise and fall to perform its office, by being mounted in a central shaft or spindle rising through the middle of the table...'.[31] For large machines the whole of the apparatus could be placed in a pit such that the table was flush with the ground. It sounded ingenious but, in fact, the machine failed to please and was progressively reduced in price.[32] Much more successful was Nasmyth's shaping machine or 'steam arm'. This was his main contribution to the development of the planing machine and may have been an improvement on a more primitive design in the Maudslay workshops. Its purpose was to plane small surfaces and detailed parts of machines, a process that had previously been done by hand. The shaper was

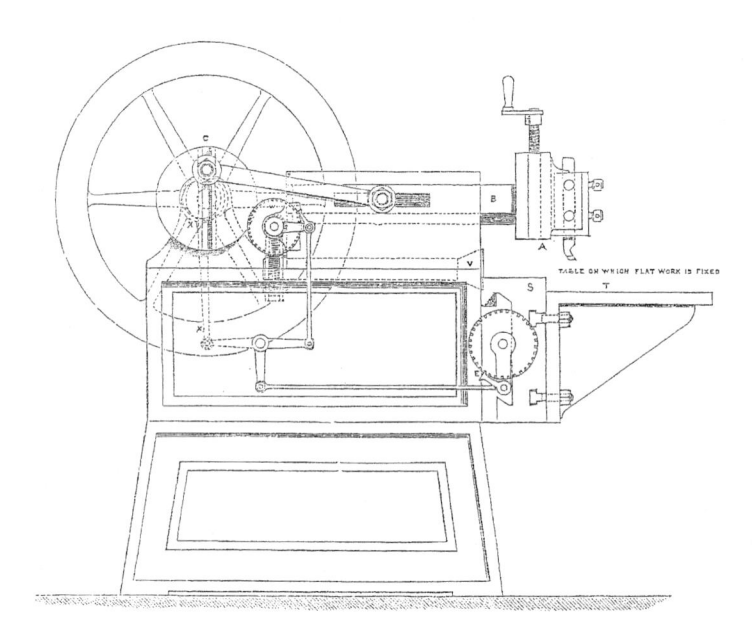

Fig.71 Nasmyth's shaping machine. *(Nasmyth, 1883)*

an undoubted commercial success at least until Whitworth's travelling-head type of shaping machine with a quick-return motion was introduced in the late 1840s. No fewer than 236 Nasmyth shapers were delivered from the Bridgewater Foundry before the end of 1856.[33]

It would appear that in the late 1830s and early 1840s Nasmyth was in the vanguard of machine tool design. Of the fifty-four machines illustrated and described in the 1841 third edition of Buchanan's *Essays on Millwork* twenty-one were designed by Nasmyth Gaskell & Co.[34] Customers for Nasmyth's machine tools for this period included many of the best known engineering establishments: Bramah & Robinson, Miller & Ravenhill, William Fairbairn, G&J Rennie, Benjamin Hick, Boulton & Watt and Maudslay, Sons & Field.[35] Yet there was virtually no change in the range of machine tools offered for sale after the 1839 catalogue. Only two new machine tools were devised after 1843, the grooving drill, for cutting the cotter holes in piston rods, and the ambidexter lathe which enabled two articles to be turned at once. The explanation for this would appear to be that since Nasmyth had invested so heavily in pattern stocks and since his order books were so full from the mid 1840s onwards with demands for steam hammers and locomotives, there was little incentive to keep on improving his machine tools especially when the old trusted designs kept on selling. Manufacturing and profit making took precedence over innovation. Nasmyth was completely unrepresented among the machine tools on display at the Great Exhibition though this was partly because, as he explained to Thomas Worthington, the pressure of urgent business meant that he would not have time to prepare all the machinery he had originally intended to send to London.[36] Perhaps he also feared being outclassed by Whitworth who, in Nasmyth's view, was overly concerned with accuracy and precision.[37] As Nasmyth explained to Charles Babbage, 'we are but rough and ready folks dealing in quantity of goods, everyday usefull sturdy work but not over refined as I do not like to boil eggs with chronometers'.[38]

In addition to steam hammers and pile drivers, steam engines and locomotives formed the other main products of Nasmyth's business. With regard to steam engines Nasmyth introduced two improvements. The first, patented in 1849, was the detached engine system.[39] This was devised to replace the expensive and cumbersome method of transmitting power to mill machinery, or machine shop tools, from a single stationary engine via a network of shafting and gearing. The new arrangement consisted of a boiler supplying steam to a large number of small steam engines attached to their respective machines. The patent applied to the method of driving machines in textile manufacture where the machines used in the delicate operations of printing, for example, were interdependent with the stoppage or variation in speed of one machine producing an effect upon the others. Nasmyth's system enabled adjustments to be carried out while the machine was working at a low velocity which could then, via a regulatory lever connected to the valve, be increased gradually and smoothly to the normal working speed. Nasmyth's detached engine system was, according to its author, 'universally applied at home as well as abroad' by the 1880s.[40] It was introduced to the Woolwich Arsenal for driving the machine tools and may also have been in use at the Bridgewater Foundry. A leading technical journal praised the system but questioned its originality since a similar arrangement had been patented for corn mills.[41]

Nasmyth's main contribution to steam engine design was his steam hammer engine. It was so called because the arrangement of cylinder, piston rod and framing was identical to that employed for the steam hammer. It was awarded a medal at the 1851 Great Exhibition and the Council noted its 'durable' workmanship and 'compact and economical' design.[42] Nasmyth also emphasised its 'get-at-ability of parts and the action of gravity on the piston', which, because it worked vertically, did not cause uneven wearing of the cylinder on one side. Nasmyth manufactured some seventy-nine of these engines, the largest being a 40hp engine for the shipbuilder John Laird of Birkenhead. The steam hammer engine soon attained a dominant position in marine engineering and was used for driving the shaft of screw-propelled steamships.[43]

Nasmyth entered the locomotive trade in 1838 when his firm signed a contract with the London & Southampton Railway Co. to supply three locomotives, the *Hawk*, *Falcon* and *Raven*. So began the association between the Bridgewater Foundry and locomotive manufacture that lasted until 1939. Between 1838 and 1853, the date of the last locomotive delivery during Nasmyth's time in charge, some 109 locomotives left the works.[44] The most important contracts were for twenty machines supplied to the Great Western Railway and another twenty machines supplied to the York, Newcastle & Berwick Railway. The locomotives produced were to the specifications of Edward Bury, Robert Stephenson, Daniel Gooch, William Norris and Archibald Sturrock. Railway contracts were of fundamental importance to the trade of Nasmyth's firm, especially during the early years, and helped to avert the worst effects of the trade recession in 1842 and 1843. Between 1837 and 1856, 37 per cent of Nasmyth's turnover related to locomotive sales.[45] During these years he patented two improvements in locomotive design. The first, dated April 1839, was for encircling locomotive bearings or journals with collars or rings of case-hardened iron or hardened steel instead of brasses.[46] The collars were to be formed in two pieces and fastened by means of a pin or key so that they were prevented from turning round upon the surface of the journal. The second patent of December 1839 was for an arrangement of self-acting brakes for railway carriages such that the carriage buffers would be connected to the brakes so that the train could be made to stop by means of its own velocity.[47] How successful these two improvements were is difficult to establish as they elicited virtually no comment in the contemporary railway journals or newspapers. They did, however, assist Nasmyth's firm in convincing prospective buyers of their competence and commitment to locomotive manufacture.[48]

Central to both Nasmyth's business career, and his engineering and popular reputation, was the steam hammer. He became almost completely personified by this machine: renaming his Penshurst retirement home 'Hammerfield'; incorporating the steam hammer into his family coat of arms which was reproduced on his memorial in Deanstone cemetery; and entitling the relevant chapter of his *Autobiography*, 'My Marriage – The Steam Hammer,' as if his invention somehow compensated for the lack of a legitimate heir.[49] The invention, protected by a patent of 1842, was the most successful machine produced by the Bridgewater Foundry. Between 1843 and 1856, 490 steam hammers left the production line at Patricroft. These generated sales revenues of more than £250,000 which represented nearly 40 per cent of his firm's trade and an even larger proportion of its profits. Custom extended across Europe to Russia and even further afield to India, Australia and South America. In Britain the Admiralty and Board of Ordnance was the major customer, purchasing sixty-one machines, while a further nine establishments including the London & North Western Railway Co. and Sharp Stewart & Co. of Manchester bought five or more machines. Nasmyth spared no efforts in promoting this machine: he made frequent visits abroad to secure orders, often attended the 'starting up' of a

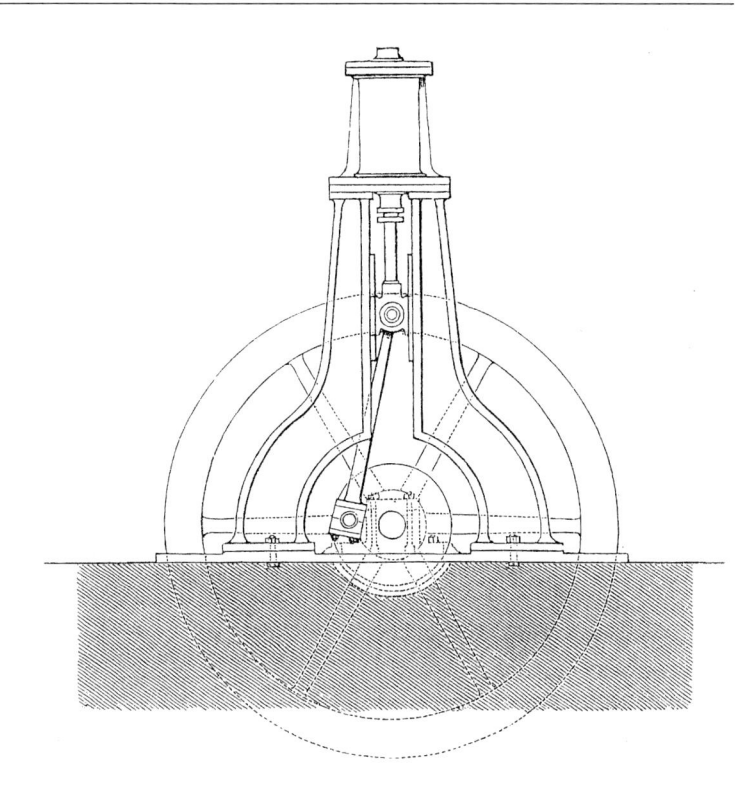

Fig.72 Nasmyth's steam hammer engine. *(Nasmyth, 1883)*

new hammer, and generally exploited every business opportunity to reach new markets. He also applied the steam hammer principle to pile driving in 1843 and made a number of important design modifications and improvements including the introduction of steam above the piston. But the truthfulness of Nasmyth's invention claim is less straightforward.

Nasmyth's own account of the invention of the steam hammer has created a legend which is still uncritically reproduced in textbooks. It was, according to Nasmyth, devised in 1839 in response to a letter from Francis Humphries, chief engineer to the Great Western Steam Ship Co. Humphries posed Nasmyth the problem of forging the massive 30in diameter paddle-shaft for the *Great Britain* which was beyond the capacity of existing helve or tilt hammers. Nasmyth tells us that he got out his Scheme Book and within half an hour of receiving the letter had sketched out his steam hammer 'in all its executant details'. The design was then shelved because the steam ship company abandoned paddles in favour of the new form of screw propulsion. In July 1840 the Bridgewater Foundry was visited by Eugene Schneider, proprietor of the Le Creusot ironworks, together with his engineer François Bourdon. Nasmyth was absent from Patricroft but the Frenchmen were shown the steam hammer drawing in the Scheme Book by Holbrook Gaskell. Nasmyth did not learn about this until April 1842 when, while in France on a business trip, he went to Le Creusot and found there a working steam hammer. Upon returning to England Nasmyth immediately applied for a steam hammer patent.[50]

In 1843 a prolonged and acrimonious dispute developed between Nasmyth and his French counterparts over the invention of the steam hammer. During the following year the

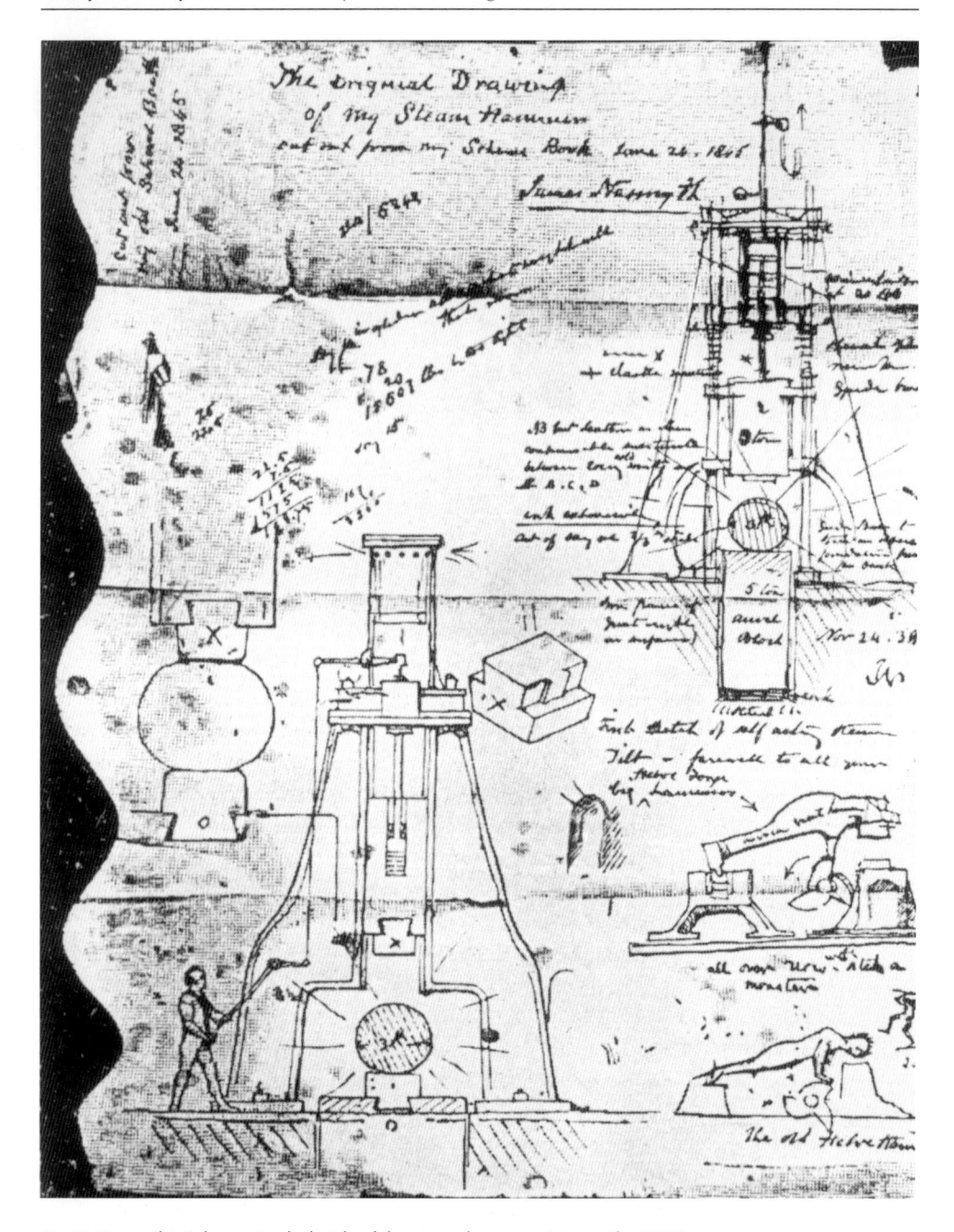

Fig.73 Nasmyth's Scheme Book sketch of the steam hammer. *(Nasmyth, 1883)*

arguments became fully public when a series of letters from the interested parties was published in the columns of the *Moniteur Industriel*, prompted by the exhibition of Bourdon's hammer at the 1844 French Exposition. Surviving records allow a number of observations to be made. First, Nasmyth's claim over the invention is not fully supported by comments and sketches made by Nasmyth in his firm's letter books. The latter firmly establish that Nasmyth had a notion of a steam hammer in November 1839 but that it was in a more primitive form from that depicted in the Scheme Book sketch dated 24 November 1839. The suggestion that the steam hammer had achieved its final form in November 1839 is not consistent with the evidence. Secondly, letters written by Schneider to his firm's commercial headquarters dated before the visit to Patricroft show that Bourdon had designed, though not built, his own steam hammer. Bourdon did not obtain the idea of a steam hammer from Nasmyth's sketch though he may have been influenced to an extent by Nasmyth's design.

The argument over priority and general principle can be settled by reference to the ideas of two British engineers who had thought of the general arrangement of the steam hammer many years before. In a patent of 1784 James Watt considered the possibility of attaching a 'hammer or stamper ... directly to the piston or piston rod of the engine'[51] while in 1806 William Deverell devoted an entire patent to what was in effect the essence of Nasmyth's later idea.[52] Nasmyth does not appear to have been aware of these patents in November 1839 though he must have seen references to them in the technical press after his patent was granted. The international controversy rumbled on with Schneider repeating the French case to a Parliamentary Committee in 1871, Nasmyth contradicting the evidence before the same committee, and articles appearing in the French technical press to forward Bourdon's claims.

This was not the only controversy surrounding the invention. Following the publication of *Industrial Biography* in 1863,[53] Robert Wilson, Nasmyth's one time works manager and then managing partner of the Bridgewater Foundry, took action to forward his own claims. Wilson maintained that it was he who had made the steam hammer a commercial proposition by devising the self-acting mechanism. The latter gave the machine operator complete command over the working action such that the hammer could break an egg in a wine glass on the anvil. Prior to Wilson's mechanism the steam hammer had failed to attract orders. Indeed when Nasmyth heard of Wilson's achievement he was so excited that he broke down Wilson's door to get an immediate look at his plans. Wilson's claims were supported by the foreman of the smiths and Holbrook Gaskell who stated that Nasmyth's own account was disfigured by exaggerations, omissions and misstatements. The full case for Wilson was published in a pamphlet entitled *History of the Steam Hammer*.[54] Interestingly, Nasmyth never publicly contradicted Wilson's claims though he privately confided to Smiles that Wilson was a conceited fellow engaging in a cowardly attempt to rob him of his inventorship.[55]

Nasmyth would certainly have acknowledged Henry Maudslay as the principal influence upon his engineering career, not least because Maudslay was a general engineer prepared to apply his engineering talents to a broad variety of work. In this sense James Nasmyth was a true disciple, for although he had a particular interest in steam engines and machine tool design, he was always prepared, providing the commercial incentives were evident, to apply his inventive skills to any engineering or engineering-related problem. Hence one of his best known improvements was in connection with foundry work, the safety foundry ladle. This simple device, consisting of a screw wheel and worm, addressed the physical danger involved when molten iron was poured into casting moulds. Its effectiveness inspired letters of commendation from the firms of Hick, Field and Whitworth.[56] Safety, this time in relation to the ventilation of coal mines, was also the focus of Nasmyth's direct action suction fan. This was successfully installed at Earl Fitzwilliam's Skyer Spring pit near Wentworth. Another patented invention involving a process for puddling iron by steam[57] clearly anticipated Bessemer's more famous patent of 1855. The range of Nasmyth's engineering interests is illustrated in the 'Chronological List of Mechanical Inventions and Technical Contrivances' which forms an appendix to his *Autobiography*. There are forty-three items ranging from floating mortars to ways of producing graceful curves on pottery and glassware, and these comprise only 'the more prominent ' of his works.[58]

It was not just as a general engineer that Nasmyth followed Maudslay. Many of Nasmyth's early machines were manufactured with the assistance of Maudslay tools and some of the most successful machines produced at the Bridgewater Foundry, such as the nut-cutting and shaping machines, had a Maudslay connection. Furthermore, Nasmyth followed Maudslay's practice of keeping a supply of popular machines in stock so that demand could be instantly satisfied. In 1836 Nasmyth was enthusing to Gaskell over the advantages of the 'ready made' concern. At that time Nasmyth saw this as the means of attaining the lead in business and a small number of machines such as wall drilling, punching and shearing, and paring and slotting machines were produced ahead of orders. Nasmyth even attempted to take the lead in locomotive construction by offering to the market ready-made engines on Stephenson's

Fig.74 Nasmyth's steam hammer showing the self-acting mechanism. (Knight, 1870)

Fig.75 The traditional and safety foundry ladle. *(Nasmyth, 1883)*

principle. In most years, however, Nasmyth's stock-selling activities were extremely limited, due to the firm's preoccupation with high-value capital goods such as steam hammers, locomotives and large stationary steam engines. It was not financially viable to keep stock items of such products and Nasmyth relied instead on maintaining a large ready-made stock of standard patterns which enabled him to engage in the production of standard advertised items. The Nasmyth catalogues were the logical extension of Maudslay's advertising leaflets of 1812 and 1813. But there are also ways in which Nasmyth did not follow in his master's footsteps. Maudslay's interest in precise measurement, standard screw threads and true plane surfaces were all areas where Whitworth took the lead, so establishing the basis of modern precision engineering. Nasmyth's interests in extending the frontiers of knowledge were in astronomy rather than engineering and once he had acquired the means to retire from business and devote himself to 'active leisure', he did so at a time when many engineers would have been at the half-way stage in their careers. That Whitworth became a knight of the realm and Nasmyth did not, perhaps reflects the latter's preoccupation with his gentlemanly pursuits.

Endnotes

[1] The best short introduction to James Nasmyth is to be found in Musson and Robinson 1969, pp.489-509. For a full length study see Cantrell 1984. There is a substantial Nasmyth archive and the main deposits can be found in Salford Archives, the British Library, the National Library of Scotland and the Public Record Office.

[2] Smiles 1883, pp.120-22, 110.

[3] *Edinburgh Journal of Science*, 6 (November-April), 1827, p.225. This apparatus worked on the principle that when a solid immersed in water was heated, the water would be displaced by the expansion of the solid. The extent of the displacement could then be measured from an index tube.

[4] *Edinburgh Journal of Science*, 6 (November-April), 1827, p.345.

[5] Various Communications by Messrs George & James Nasmyth to The Society of Arts for 1826, 1827 and 1828: NLS Acc 4537/8. Nasmyth's application of waste steam constituted an independent discovery of the principle of steam blast first used by Trevithick in his Pen-y-Daren locomotive but credited by Smiles to George Stephenson – see Rolt 1960, p.40. George Nasmyth (1806-c.1864) trained and worked alongside his brother until 1843; James Nasmyth expunged his brother from the record as a result of George being disgraced in 1859 when he was dismissed from his post as curator of the Patent Office Museum for embezzlement of public funds: see Cantrell 2002.

[6] Report of the Prize Committee of the Society of Arts for Scotland, 3 June 1829, NLS Acc 4534/11. The improved method was described to the Society on 5 December 1826. The Report states in reference to the Nasmyth brothers 'The numerous communications which these gentlemen have made, and the zeal which they have evinced in pursuing their mechanical researches, merit great praise.'

[7] Smiles 1883, p.124.

[8] Cooksey 1991.

[9] Smiles 1883, pp.79-82 & 91. The Library Subscription records for the Royal High School, Edinburgh, indicate that the Nasmyth brothers attended the school between 1818 and 1822: ECA SL137/15/4 and SL137/15/5.

[10] Smiles 1883, pp.93-96.
[11] The Edinburgh University Matriculation Albums show that the Nasmyth brothers enrolled in 1823, 1825 and 1826.
[12] Smiles 1883, pp.117-119, 122-123.
[13] For Nasmyth's work as an astronomer see Chapman 1996 and 1998.
[14] Smiles 1883, pp.100-101.
[15] Smiles 1883, pp.405-408.
[16] J. Nasmyth to D.O. Hill, 5 August 1845: ROE Archives A9. 4ZB.
[17] A. Nasmyth to T.W. Winstanley, 19 July 1832: MPL, M6/1/53/75-76.
[18] This was the senior Nasmyth's first term for what became known as the bowstring bridge, a structure formed of an arch of wood or iron, often braced, the thrust of which is resisted by a tie forming a chord of the arch.
[19] See p.35, note 1.
[20] Smiles 1883, pp.139-178, 408-409.
[21] Smiles 1883, p.178.
[22] Nasmyth retired with £110,000.
[23] Cantrell 1981.
[24] Cantrell 1985.
[25] J. Nasmyth to D.O. Hill, postmarked 5 August 1835: ROE Archives A9. 4ZB.
[26] Cantrell 1984, p.39.
[27] Cantrell 1984, p.67.
[28] Smiles 1883, p.190.
[29] J. Nasmyth to P. Ewart, 24 May 1838: SA, Letter Book 2, p.167.
[30] See p.60.
[31] Patent No.7815, 1838.
[32] Cantrell 1984, pp.100-105.
[33] Cantrell 1984, pp.105-107.
[34] Buchanan 1841, pp.427-69.
[35] Between 1843 and 1855 eight machine tools were delivered from Patricroft to Lambeth including two great vertical paring machines, two punching machines and two planing machines. Sales to Maudslays during these years amounted to £3,943.
[36] J. Nasmyth to T. Worthington, 7 March 1851: MPL M6/3/11/251.
[37] Nasmyth's views on Whitworth's measuring apparatus capable of detecting a millionth part of an inch are made plain in the form of a note, entitled 'Nasmyth', written in the hand of Smiles and located among some Nasmyth/Smiles correspondence. It would appear to be Smiles' record of a conversation with Nasmyth. Nasmyth is described as having made sarcastic remarks about Whitworth's machine claiming that it was so delicate that were two persons to look at it at the same time, the device would immediately be put out of order: BL, Add. Mss. 71076 f.150.
[38] J. Nasmyth to C. Babbage, 17 January 1851: BL, Add. Mss. 37194 f.474.
[39] Patent 12675, 1849.
[40] Smiles 1883, pp.316-17.
[41] *Practical Mechanics Journal*, 3, 1851-52, p.142.
[42] Smiles 1883, pp.167-68.
[43] Dickinson 1939, p.144.

[44] Cantrell 1984, pp.264-67.
[45] Cantrell 1984, p.217.
[46] Patent 8023, 1839.
[47] Patent 8299, 1839.
[48] Nasmyth also took out a patent in 1844 in connection with atmospheric railways. See Cantrell 1984, pp.120-21.
[49] Nasmyth fathered a daughter, Minnie, by his mistress Virtue Squibb who assumed the name of Emily Russell between the 1860s and early 1880s. Letters between Nasmyth and Virtue, whom he addressed as Floss, have recently been deposited with the National Library of Scotland by Nasmyth's great-grandson, the late Mr John Russell Abbott, and his great great-grandson, Mr Christopher Russell Abbott.
[50] Patent 9382, 1842.
[51] Patent 1432, 1784.
[52] Patent 2939, 1806.
[53] Smiles 1863.
[54] Rowlandson 1864.
[55] J. Nasmyth to S. Smiles, 6 August 1882: BL, Add. Mss. 71076 ff.155-156.
[56] NLS, Acc 4534/111.
[57] Patent 1001, 1854.
[58] Smiles 1883, pp.400-39.

9
William Muir
Tim Procter

William Muir (1806-1888) is one of the less well-known members of the Maudslay school.[1] Unlike Maudslay, Whitworth, Nasmyth or Roberts, Muir has no singular development to his name, nor a reputation for inventive flair. While Samuel Smiles could take for granted that the readership of *Industrial Biography* would know who Muir was in 1863,[2] Rolt did not refer to him at all in 1965, while other twentieth-century engineering historians mention only that he went on to work for Joseph Whitworth.[3]

Economic studies of the machine tool industry have placed the Muir name in a more dynamic light. The firm he founded in Manchester, William Muir & Co., is described by Floud as 'among the most reputable and most successful makers of machine tools at this period...'[4] while Saul says that Muir & Co., along with J.S. Hulse & Co. and Cravens, 'formed the basis upon which a new generation of specialised heavy machine tool makers grew up in Manchester'.[5] Muir links the pioneering period of the Maudslay school to the mature stage of the machine tool industry in the late nineteenth and early twentieth centuries. His firm gained a reputation for high quality products and maintained its position as a market leader alongside newer manufacturers with equally strong reputations. Arguably this makes him one

Fig.76 Portrait of William Muir, date unknown. *(From the collections of the Museum of Science and Industry in Manchester)*

of the most successful of the Maudslay school and worthy of the same renown enjoyed by some of his peers.

William Muir was born in Catrine in Ayrshire on 17 January 1806, the third child of Andrew and Agnes Muir. He was baptised on 26 January. His father appears to have been a tenant farmer who also worked as a carrier or carter between Catrine and Glasgow, and his family were 'worthy people of the middle class, in comfortable circumstances'.[6] William was one of seven children born to Andrew and Agnes between 1802 and 1819. His father was still living in Catrine in 1851, aged seventy-four, working as a merchant grocer.

According to Robert Smiles, Muir became interested in engineering while at school, and entered a five-and-a-half-year apprenticeship with Thomas Morton of Kilmarnock. Morton's main business was repairing carpet-weaving looms, but he also patented several improvements to the loom, made telescopes and developed a screw propeller. Morton had a profound influence on Muir. Writing in 1884, Muir described Morton as setting him 'an excellent example alike in the house, in the workshop, and in the world ... He had to contrive many of his peculiar tools and appliances, to realise results that existed in some instances only in his own mind ... It was a great advantage to me to have a master of such an active, searching, thoughtful, enterprising spirit as inventor and improver; I could scarcely fail to catch a spark from such a fire'.[7] Morton powered the lathes in his works with a windmill; this was probably Muir's first exposure to a large-scale application of independently powered machine tools, and possibly inspired his interest in machine tools and the lathe in particular. Some of Muir's work echoes Morton's interests. Morton made optical instruments; Muir patented a theodolite with Henry Goss. Morton experimented with screw propellers; Muir worked on Bennet Woodcroft's patent screw propellers in the 1840s.

In 1824 Muir joined Girdwood & Co. of Glasgow, makers of cotton-spinning machinery. He moved to the Catrine Cotton Co. in 1829, and then spent a short time with Henry Houldsworth of Glasgow. Robert Smiles notes that while at Houldsworth's, Muir became exasperated with the 'old jigger' of a lathe that he had to work on, and that the study, development and perfection of the lathe was 'almost a passion' for him.

Muir left Glasgow in September 1830, travelling to Liverpool and then to Manchester, where he intended to stay. However he went to Truro on hearing that his brother Andrew was ill, and remained in Cornwall, joining Harvey at the Hayle Foundry. A considerable amount of Harvey's business came from supplying steam engines for pumping water out of copper and arsenic mines, and it is almost certain that some of Muir's work for them would have been on engines or engine parts. Around March 1831 he left Harvey, for reasons unknown, went to London, and joined Maudslay, Sons & Field on 25 April.

Henry Maudslay died on 15 February 1831, just before Muir joined his firm. Muir therefore cannot be called a true pupil of Henry Maudslay, but the spirit in which Maudslay ran the business was very much carried on by his sons, Thomas Henry and Joseph, and by Joshua Field. Muir came to the firm with broad engineering experience. He was also interested in improving the machine tools he used, if Robert Smiles' story about the lathe at Houldsworth's is true. Therefore the atmosphere and style of Maudslay's is likely to have suited Muir well.

Muir became a foreman at Maudslay's, but in the absence of his own papers or any contemporary records of Maudslay, Sons & Field, it is impossible to say for certain exactly what his job was in the five years he spent there. Robert Smiles' description suggests that he was a general foreman, supervising construction and keeping track of men's time on particular projects, rather than being in charge of a single department.[8] While marine steam engines formed the greater part of the firm's business at this time, it also undertook a variety of other contracts. These included pumping engines for the Lambeth and Grand Junction Water Works, the time ball for the Greenwich observatory, and the iron ship *Lord William Bentinck*. Robert Smiles mentions Muir working on a valve for the Lambeth engine, a new drilling machine for the factory, the 'Nantz 30 h.p. vibrating steam engine' and a steam carriage designed by Admiral Thomas Cochrane. From this frustratingly brief description it is clear that Muir worked in a lot of different areas, including machine tools. Robert Smiles also mentions that while there, Muir met Joseph Whitworth and James Nasmyth.[9]

In March 1836, Muir left Maudslay's, again for unknown reasons. His time there encouraged and developed his mechanical interests and skills – five years in a senior position at Maudslay, Sons & Field was bound to nurture anyone with talent in this field. The products of Muir's own firm were most admired for their quality, and it is tempting to say that high quality was an ideal instilled in him during his time in the Maudslay school. However it appears that, unlike many of the other Maudslay school graduates, he did not leave with the intention of setting up on his own.[10] After leaving Maudslay's, he next spent just under nine months with Charles Holtzapffel & Co. at Long Acre, as a representative travelling all over the country, and then over three and a half years with Bramah & Robinson as a foreman at their Pimlico works. Both these firms had reputations for excellence almost as formidable as Maudslay's. Muir had now held senior posts with three of London's most renowned and respected engineering firms, and had he been contemplating setting up on his own, this would have been an ideal time to do it.

However, at the end of June 1840, Joseph Whitworth recruited Muir as a manager at his Chorlton Street works in Manchester. The two men may not have met at Maudslay's, but it is inconceivable that Whitworth would employ Muir if they had not met or got to know something of each other in London. Muir worked, to a greater or lesser degree, on some of Whitworth's most well-known endeavours. Robert Smiles looked over Muir's notebooks and sketch books for his memoir, and picked out drawings and calculations for the road-sweeping machine and various experiments on screw threads. He also suggested that Muir deserved more credit for the work he did on Whitworth's machine tools, including the radial drill. Their parting company was probably due to differences in personality, religious beliefs, and attitude to mechanical design.[11] Muir left Whitworth's in June 1842, and finally went into business on his own, seven years after leaving Maudslay.

Muir's first premises were in Berwick Street, Manchester, where he had a small forge, his lathe, and work benches. This was typical of the way most of the Maudslay men started out, in small workshops, often on their own. However Muir quickly hired assistance. David Mackley, a pattern maker, joined Muir in 1842 and may well have been his first employee. William D. Parks, a smith, joined him in 1846. These two were still working for William Muir & Co. in December 1883, when they signed an address to Muir, so there may well have been others who had worked for Muir in the early 1840s and who had since retired.

Muir's first major commission was the manufacture of railway ticket apparatus invented by Thomas Edmondsen. Edmondsen's system consisted of ticket printing, dating and counting

Fig.77 Muir's advertisement of 1850. *(Slater's Manchester & Salford Directory, 1850)*

machines, and it had attracted considerable interest from railway companies. Edmondsen was not a metal worker and he had come to Manchester to be closer to potential manufacturers of his machines. He had approached Whitworth just before Muir left the Chorlton Street works, but Whitworth was 'unwilling to be bothered with what he probably considered pottering work'[12] and passed Edmondsen on to Muir. Once Muir left Whitworth's, his association with Edmondsen appears to have been very fruitful, as the two men were eventually to move their businesses into the same premises. Curiously, while Muir did not have an aversion to publicising designs made in conjunction with others, he never mentioned his work on Edmondsen's apparatus in his advertising, or displayed it at the exhibitions he attended.

By 1845 Muir had moved to 59 Oxford Street, and an advertisement he placed in Slater's *General & Classified Directory of Manchester & Salford* in that year gives a good idea of the scale of his business:

> *To Amateur Turners, Patentees, Inventors & Others: William Muir (from London), No.59 Oxford Street, opposite the Atlas Works, Manchester, Machinist, Tool & Model Maker and Manufacturer of lathes, tools, improved copying presses, and machines of every description, for ornamental or useful purposes.*
>
> *Amateurs, inventors etc. can be supplied with the apparatus, tools and materials required for turning and the mechanical arts generally, and for making models etc., and can be practically instructed in their use, for the various branches of the art, on reasonable terms.*

Muir was now a manufacturer of machine tools, albeit small and bench top types. However he was also working on larger machines. Holtzapffel described a planing machine for floor boards built by Muir which performed eight simultaneous operations on deal boards fresh from the saw mill, and which left them 'in a condition nearly ready for fixing ... but sometimes a little finishing with the hand-smoothing plane, is required at those parts where the grain is unfavourable to smooth cutting'.[13]

The letter-copying press mentioned in the advertisement was also his own design, and was to prove one of his most enduring products. William Muir & Co. would stop making it only after his death, long after they had ceased to manufacture many of the other machines that Muir first sold. Around this time Muir also worked on Bennet Woodcroft's screw propellers and improved revolving spiral paddles, patented in 1844 and 1846 respectively. Soon after this he moved again, this time with Edmondsen, into much larger premises on Miller's Lane, Greengate, Salford.[14] These were in part of the former Salford Iron Works of Bateman & Sherratt. Edmondsen's ticket printing office occupied the upper floor, and Muir used the main portion of the works for manufacturing Edmondsen's apparatus and his own expanding product line. He was on the look-out for new machinery at this period, as on 28 March 1850 he wrote to James Watt & Co., steam engine makers and engineers of Soho Foundry, near Birmingham, about a forthcoming auction at their Soho Manufactory works which were about to close.[15] Muir took out an advertisement in Slater's *Directory of Manchester and Salford* in 1850, which illustrates how both his product range and his connections with other Manchester engineers and inventors were widening. He was making oil testing machines invented by Emmanuel Thomas of Manchester, and coffee mills invented by William Weild, of Sharp Stewart & Co., who was to propose Muir for membership of the Institution of Mechanical Engineers in 1863.[16]

In 1851 Muir exhibited a selection of his products at the Great Exhibition. These included the copying and embossing presses, a foot lathe and slide rest, Weild's coffee mills, Thomas' oil testing machine, a soap cutting machine invented by Walter Storey, and the theodolite which he patented in the same year with Henry Goss, Joseph Whitworth's head draughtsman. Muir had stated in his advertisement of 1850 that he had 'not secured any Articles of his Manufacture by Patent or otherwise...' but he would from now on take out several patents for his developments. His increased patenting activity may well have been on Goss's advice.[17]

The 1850 advertisement and the Great Exhibition provide a snapshot of Muir just as his career and business were about to change. He had a workforce at Miller's Lane – the advertisement says that products were made under 'his own immediate direction and superintendence' – but he was not yet making heavy machine tools. However the insistence on high quality was already well-established. The advertisement insisted that Muir depended 'for his success upon first-rate quality...', and the tools he exhibited at the Great Exhibition earned him a prize medal.

In the space of three years Muir's business greatly changed. An advertisement in the *Manchester Mercantile & Manufacturing Annual Directory* for 1854-55 proudly described the products of 'Wm. Muir & Co., Engineers, Tool Makers, Machinists, and Iron Founders' of the Britannia Works, Sherborne Street, Strangeways, Manchester. The copying presses were still doing well – 'The increasing demand for these Presses is a sufficient guarantee of their excellence' – and they were being sold through a London agency, T.J.& J. Smith, wholesale stationers of 83 Queen Street, Cheapside, London. However the rest of the product range was now taken up almost entirely by machine tools: bolt-screwing machines, foot and power lathes of all sizes, self-acting drilling, slotting and planing machines, shaping machines, punching and shearing machines, nut-shaping machines and wheel cutters. The firm also opened a London office during this period, situated first at 3 Victoria Street, Westminster, and later at 10 John Street, Adelphi.

Clearly Muir's manufacturing capacity had increased with the move to the new works. Britannia Works was just within Manchester, around a quarter of a mile north of Muir's former premises in Salford. It was situated on one-and-a-half acres of land on the banks of the River Irwell, leased from the Howard family, the Earls of Ducie. It was built in 1851 and William Muir & Co. began full-scale manufacturing there in late 1851 or early 1852. Expansion to large-scale works and heavy manufacturing was sometimes dependent on attracting partners who could provide financial backing. Of the Maudslay men who went down this route, some, such as Nasmyth, found backers relatively quickly while others, such as Francis Lewis, took longer to find willing investors. However it is possible that this was not the case with Muir. According to Robert Smiles, Muir took out the lease and designed the works on his own account and only then did he take partners to establish William Muir & Co.[18] Self-financing such a move would have required a steady supply of quite large or lucrative orders, but all the large-scale contracts mentioned in extant sources date from after the opening of Britannia Works. In the absence of the firm's books it is impossible to say what scale of work Muir was executing at Miller's Lane.

One possible scenario is that Muir secured financial help to build Britannia Works, and then took his backers into partnership, but details of who they were have not survived. He

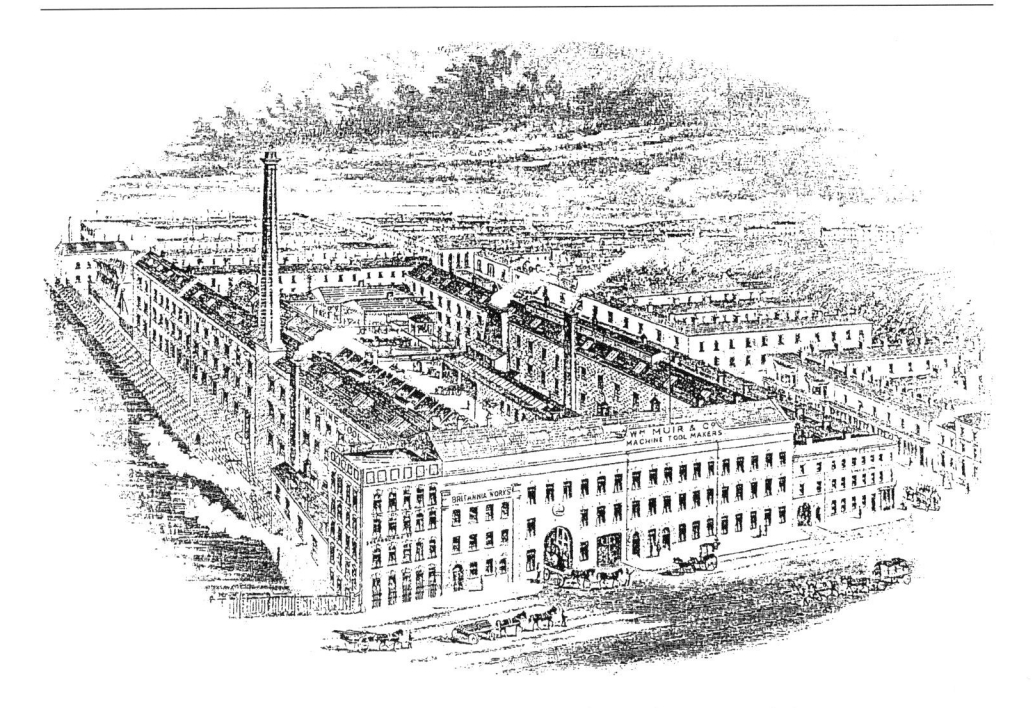

Fig.78 Britannia Works, c.1895. *(Catalogue of W. Muir & Co.)*

would have been seen as a relatively reliable investment in a lucrative field. By the early 1850s Manchester was established as a centre of engineering excellence, and mechanised manufacturing and the railways were rapidly growing. Such factors made a heavy machine tool maker a promising investment, and Whitworth & Co., Nasmyth Gaskell & Co. and Sharp Stewart & Co. were local examples of such successful speculation. Given the fact that in 1878 his son Alfred was running the firm in partnership with two of the Garnett family, who had invested in Nasmyth Gaskell & Co. in 1856, it is tempting to believe that he did have financial assistance, but no concrete information on any backers has survived. Industrial histories written in the late nineteenth century played down or totally omitted the financial and administrative assistance that well-regarded engineers received. Samuel Smiles says nothing of it in *Industrial Biography*, while James Nasmyth did not mention any of his patrons and backers in his *Autobiography*.[19] Robert Smiles' portrayal of Muir as possessing 'self-dependence and originality' is typical of the way in which successful engineers were portrayed, so perhaps it is not surprising that any helping hand remains unrecorded.

However it is possible that Robert Smiles was honest in his description of Muir's independence. While he certainly did take on partners in the early 1850s, they may not have been men who invested heavily in the firm. From an analysis of the books, Saul described William Muir & Co. in 1867 as still being small, with an output of £12,350 and a workforce of around a hundred.[20] It is not inconceivable that the business grew to this size without large injections of external capital. Moreover much of the growth of Britannia Works took place long after this time. For example large extensions to some of the shops were noted as being built in 1887 when the Iron and Steel Institute held their autumn meeting in Manchester, while Robert Smiles, writing in 1888, said that the works had 'been extended from time to time to what is now, and has long been known as the "Britannia Works".' A visit by the Institution of Mechanical Engineers in July 1894 noted that the works had gradually expanded over the years and was only then occupying all the available ground. So Muir's partnership in the early 1850s may have been a technical one with other engineers. It might have included his eldest his son Andrew, who was closely involved in the firm's design and manufacture of small-arms-making machinery in the 1850s.

In 1867 the firm became a limited liability concern.[21] William Muir & Co. Ltd was incorporated on 7 January 1867, with a potential capital of £75,000 in 15,000 shares of £5 each. At registration only 1,415 shares had been issued. Muir himself held 1,000 of these and his sons Andrew and Edwin held 100 each. The next largest allocations were three blocks of twenty shares, to Thomas Dunn, a mechanical and civil engineer of Pendleton, and to John Moore Smith and John Turk Lacey, two London architects. The next largest shareholders, with ten shares each, were Joseph Cooper, an engineer and machinist of Hulme, and Henry Bernouilli Barlow, a Manchester patent agent. The rest of the issued shares were held by various Britannia Works employees, including clerks, pattern-makers, smiths and turners, in blocks ranging from one to five. This share-holding arrangement might have been an experiment by Muir, to tie his sons and key workers into the firm. However it may not have ever formally operated, despite being registered. Late 1866 to early 1867 was the time at which Muir left Manchester for London, relinquishing his active role in running Britannia Works. Moreover his fourth son Alfred, who was to take over the running of the firm, is absent from the list of shareholders. In fact Alfred appears to have been unaware of the arrangement. In July 1878 the Registrar of Companies wrote to the firm concerning William Muir's continued involvement. The firm replied on 31 July:

> *Mr. William Muir left this firm about the year 1866 and the present partners are Alfred Muir, Charles Garnett, and Robert Garnett, and so far as they are concerned do not know anything of the William Muir & Co. Ltd.*

The assertion that the firm had never been a limited concern was repeated on 13 October 1882 in response to another enquiry from the Registrar of Companies. On paper at least the firm had been a limited concern, and the Certificate of Dissolution for this incarnation of William Muir & Co. Ltd is dated 31 August 1883.

With the opening of Britannia Works, Muir was able to develop a wide range of heavy machine tools, as described in the 1854 advertisement. The firm usually exhibited at the numerous international exhibitions, which enhanced their reputation considerably. At the 1855 *Exposition Universelle* in Paris, a small foot lathe, a vertical drill and Muir's patented double grindstone came in for singular praise. The lathe was described as 'bien disposé et d'une exécution remarquable', while the jurors concluded that 'les outils de M. Muir sont fort ingénieux, bien disposés et bien exécutés; quelques dispositions remarquables lui sont tout à fait particulières. Les machine-outils qui sortent de cet atelier sont fort estimés en Angleterre'.[22] High praise indeed. Muir's tools were awarded a 2nd class medal, alongside Whitworth who won a Grand Medal of Honour for a range of fifteen tools, and Smith Beacock & Tannett of Leeds whose nine machine tools earned them a 1st class medal.

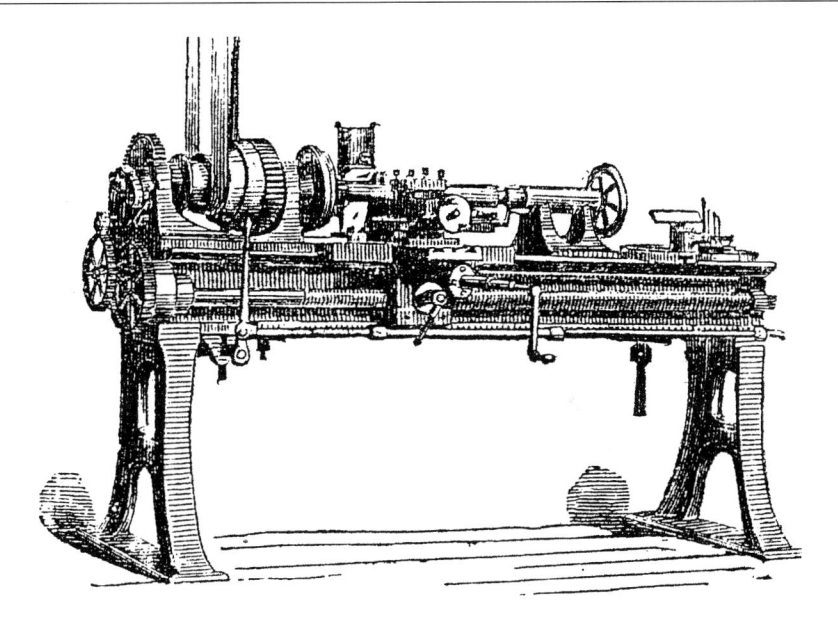

Fig.79 Muir's patent 7in centre double-geared lathe. *(1862 Exhibition Catalogue)*

Muir and most of his peers, both those of the Maudslay school and newer firms, exhibited extensively at the 1862 International Exhibition. The catalogues and reports give a good idea of his range of tools and the impression they made. Among the machine tools he exhibited were a self-acting duplex lathe with a 12in centre and 25ft bed, for sliding and screw cutting; a double-geared lathe with a 7in centre and 6ft bed, again for sliding and screw cutting; an 8in foot lathe with a double treadle, designed for use on board sea vessels or in areas of the colonies where labour was so cheap that steam power was not required; a self-acting radial drilling machine with a vertical elevating slide radial arm, similar to Whitworth's radial drill; a self-acting vertical double-geared drilling machine; a self-acting universal shaping machine capable of planing an object 2ft long or doing circular work on objects of 12in in diameter; a self-acting slotting and shaping machine able to accommodate wheels up to 3ft in diameter; the patent double grindstone which had been exhibited at Paris; smaller tools and accessories such as a bench-top drilling machine and a small planing machine, both of which could be worked by hand or an overhead power source; taps and dies; and the ubiquitous copying press.

The International Jury which judged the machine tool class included Joseph Whitworth, and their report was written by Maj. Conti of the Italian Corps of Military Engineers. The collection of tools on display comprised 1,800 machines from 442 exhibitors. The jury felt that it surpassed the collections shown in 1851 and 1855, but that it was less impressive in terms of new inventions and improvements. Such developments, the jurors noted, 'are only made when sheer necessity urges'.[23] Necessity was not urging in 1862. The pioneering phase of the machine tool industry, as represented by Maudslay and Whitworth's developments, was over. Machine tool makers now had to supply high-quality, reliable, accurate tools, incorporating minor improvements to increase their efficiency. Muir's tools were ideally suited to these market conditions, and the jury's comments reflected this. They singled out lathes as being very well represented but lacking in innovation, the sole exception being Smith Beacock & Tannett's special adaptations. However Muir's duplex lathe was noted for the long thread screw which rapidly withdrew the cutting tool when turned by a handle. His radial drilling machine was mentioned only as employing a Whitworth-style arm; Muir had worked on Whitworth's radial drill designs during his time at Chorlton Street. Of his vertical drill, which was similar to the machine he had exhibited in Paris in 1855, they noted that he had preserved the high frame of previous designs, but that he had adopted a rack-like spindle end moved by a pinion, a movement which was also used by Fox. They also noted that the drill had only vertical and circular motion. The shaping machine was described as being of the 'common pattern', but it was singled out with similar machines made by Shepherd & Hill and Zimmerman as being of accurate workmanship. The slotting and shaping machine was simply described as being 'of the ordinary kind' but the planing machine was rewarded for 'good workmanship and economical construction'. Overall the jury was impressed enough to give Muir's range of machine tools one of the twenty-six medals awarded to machines working in metal. Generally these were won by ranges of good quality tools, for example Napier & Son's printing and bullet-making machines, Sharp Stewart & Co.'s machine tools and Whitworth's small-arms work. Whitworth's machine tools would have earned him another medal had he not been a juror for that class.

This is not to say that Muir's tools had no innovative features. In 1853 he patented a quick-release motion and various other features for screw-cutting lathes; in 1856 he produced a design for adding additional slide rests to standard slide lathes in order to economise time and labour; and his double treadle mechanism was patented in 1860. These features could be seen on the machines exhibited in 1862. One of his patents had distinct echoes of Maudslay and Whitworth's work on flat metal surfaces. This was a patent taken out in 1867 for making sliding surfaces, such as the beds of planing machines or the slides of steam engines. Muir proposed making such surfaces with recesses in them and filling the recesses with block tin

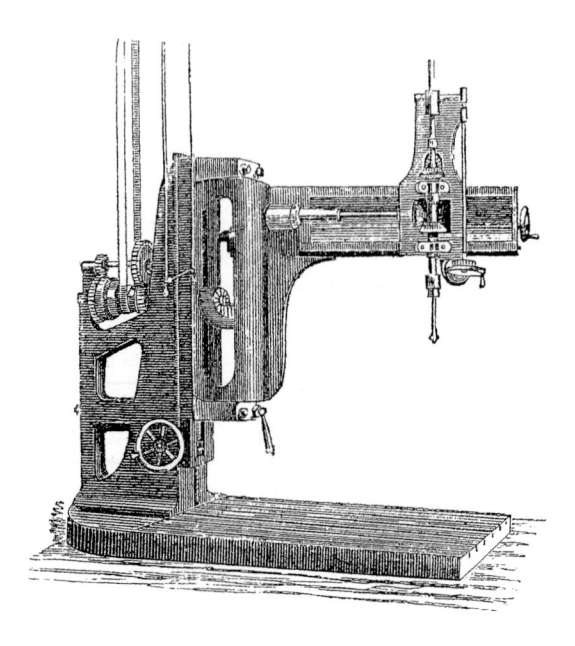

Fig.80 Muir's self-acting radial drilling machine. *(1862 Exhibition Catalogue)*

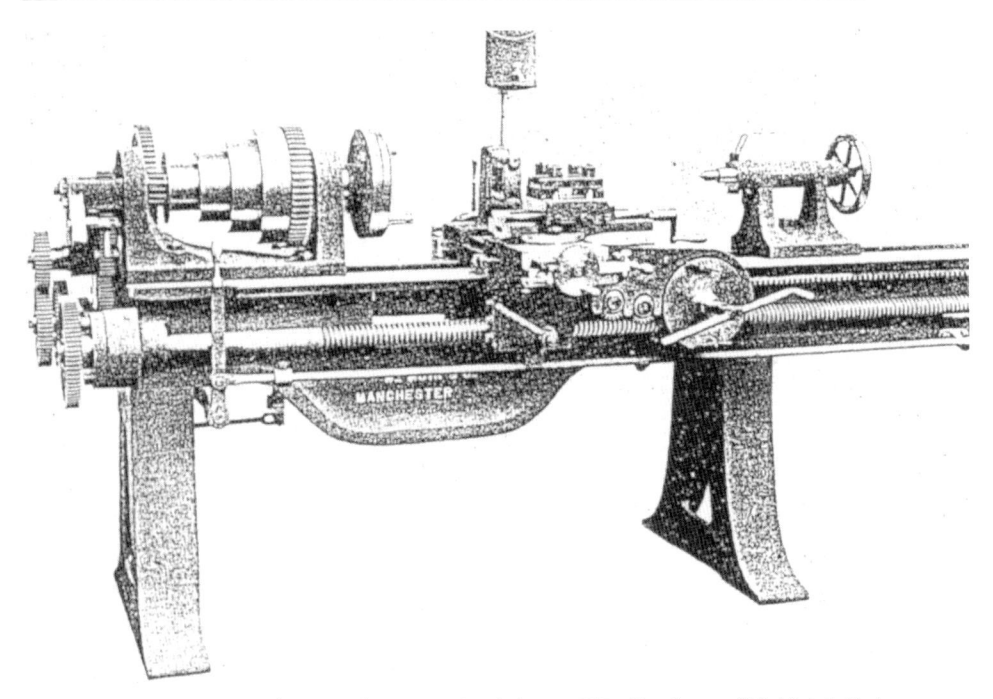

Fig.81 Muir's self-acting, surfacing and screw-cutting lathe, c.1895. *(Catalogue of W. Muir & Co.)*

or 'other antifriction metal', which would reduce the wear on the sliding surfaces. This patent received only provisional status and Muir does not appear to have used it.

Muir's machine tools were not revolutionary. Although many of his developments were extremely practical, in some areas he favoured features which were becoming outmoded, such as rack and pinion assemblies. This was something that he and Whitworth almost certainly disagreed about. It was quality and reliability that made Muir's name, not constant evolution. A comparison between the tools exhibited in London in 1862 and those advertised around 1895 shows that very little had changed in over thirty years from Muir's initial designs.

With no sales books or incoming order books to go on, little evidence remains as to who actually bought Muir's tools, or which were the most popular. The surviving records of other firms give some clues. The Manchester locomotive manufacturers Beyer Peacock & Co. bought several Muir tools in the 1850s. These included a power lathe with a 10in centre and 16ft bed, with driving apparatus, valued at £175; an 8in centre slide lathe with an 8ft bed, featuring Clement's and common drivers, value £125; a 6in centre slide lathe again with Clement's and common drivers, worth £85; a double-geared upright drilling machine with driving apparatus, £60; and a slotting machine with top driving apparatus, again worth £60.[24]

The steam engine manufacturing and engineering firm James Watt & Co., of Soho Foundry in Smethwick, near Birmingham, which developed out of Matthew Boulton's and James Watt's original partnership, made several enquiries about Muir's products in the 1860s. James Watt & Co. carried out a large number of overseas contracts, and would often supply other machinery both for the maintenance of their steam engines and the works they would be used in. They bought items they did not make themselves, such as machine tools, from other firms.

On 18 January 1864 they asked William Muir & Co. to quote for a drilling machine to bore 3in holes, a shaping machine with 10in stroke suitable for straight or circular work, a planing machine able to accommodate pieces up to 3ft in diameter and 14ft long, rollers and frames for building boiler plates, a small hand lathe, a screwing machine with a set of Whitworth standard taps and dies from 2in to ¼in in diameter, and a further set of Whitworth standard taps, stocks, dies and wrenches from 2in to ⅛in diameter. This enquiry, for an arsenal at Poonah in India, came with the proviso that no part could weigh more than 360lbs or exceed 12ft in length, as they had to be carried by mules over mountainous terrain. This order appears not to have been placed, but in May 1864 James Watt & Co. ordered a double-geared hand lathe with a 6in centre and 4ft bed, equipped with Muir's patent treadle and slide rest arrangement. The letter noted that the price would be £44, minus a commission of 5 per cent and with a further 2½ per cent reduction for payment by cash. In January 1865 Watt & Co. ordered, for another overseas client, a 12in screw-cutting lathe with a 30ft bed and Clement's and common drivers priced £253, a portable vertical drilling machine to be worked by treadle priced £32, and a portable punching and shearing machine to punch holes of ¾in diameter through plate ½in thick, priced £50. A screw- and nut-tapping machine included in the original enquiry of 17 October 1864 was not taken up. Yet another enquiry was made on behalf on an overseas correspondent in June 1868, this time for a large 22in centre lathe with a 20ft bed and a die-turning lathe with an 8ft bed. Muir & Co. duly submitted their estimate but were informed on 30 July 1868 that Watt & Co. had been told by their London agents that 'the cost of the lathes is so much more than our correspondents anticipated that they have decided not to order them'. Such baulking by clients did not stop Watt & Co. coming back to Muir's again. In July 1871 they ordered a universal shaping machine, £77 4s, with a parallel vice, £13 16s extra, which appears to have been for their own works at Soho Foundry. In November 1871 they ordered another portable punching and shearing machine to punch holes of ½-¼in diameter and to shear plates up to 12in from the edge, while on 15 October 1872 they enquired whether Muir's could supply a drilling machine for their boiler department to drill four to six holes of ⅜-1in in diameter simultaneously.[25]

In a pleasing turn of fate, Muir & Co. sold machine tools to their founder's mentors. When Maudslay, Sons & Field closed and sold off the plant at the Lambeth Works in 1900, lot 902 in the Turnery was a Muir double screw-cutting lathe with a 10in centre and 12ft bed, and lot 1172 in the Lower Vice Loft was a Muir profiling and milling machine, while at the sale at the Greenwich works two years later, lot 343 was a Muir double-geared radial drilling and shaping machine.

William Muir & Co.'s success was not founded solely on machine tools. As with many similar firms, they widened their product range as their business and premises expanded. Two long-lived products of Muir's own design have already been mentioned, the letter copying and embossing presses and the double grindstone.

Muir continued to devote considerable attention to copying and embossing presses. In 1858 he patented a stand for presses and other small machines, which he exhibited along with his improved press in 1862. In 1864 he took out a further patent for improved foundation plates and beams for copying presses and for a machine for bending the beams into the correct shape. Saul estimates that in the 1860s presses accounted for 17 per cent of the firm's output[26], and indeed in 1871 William Muir & Co. were still describing themselves as manufacturers of

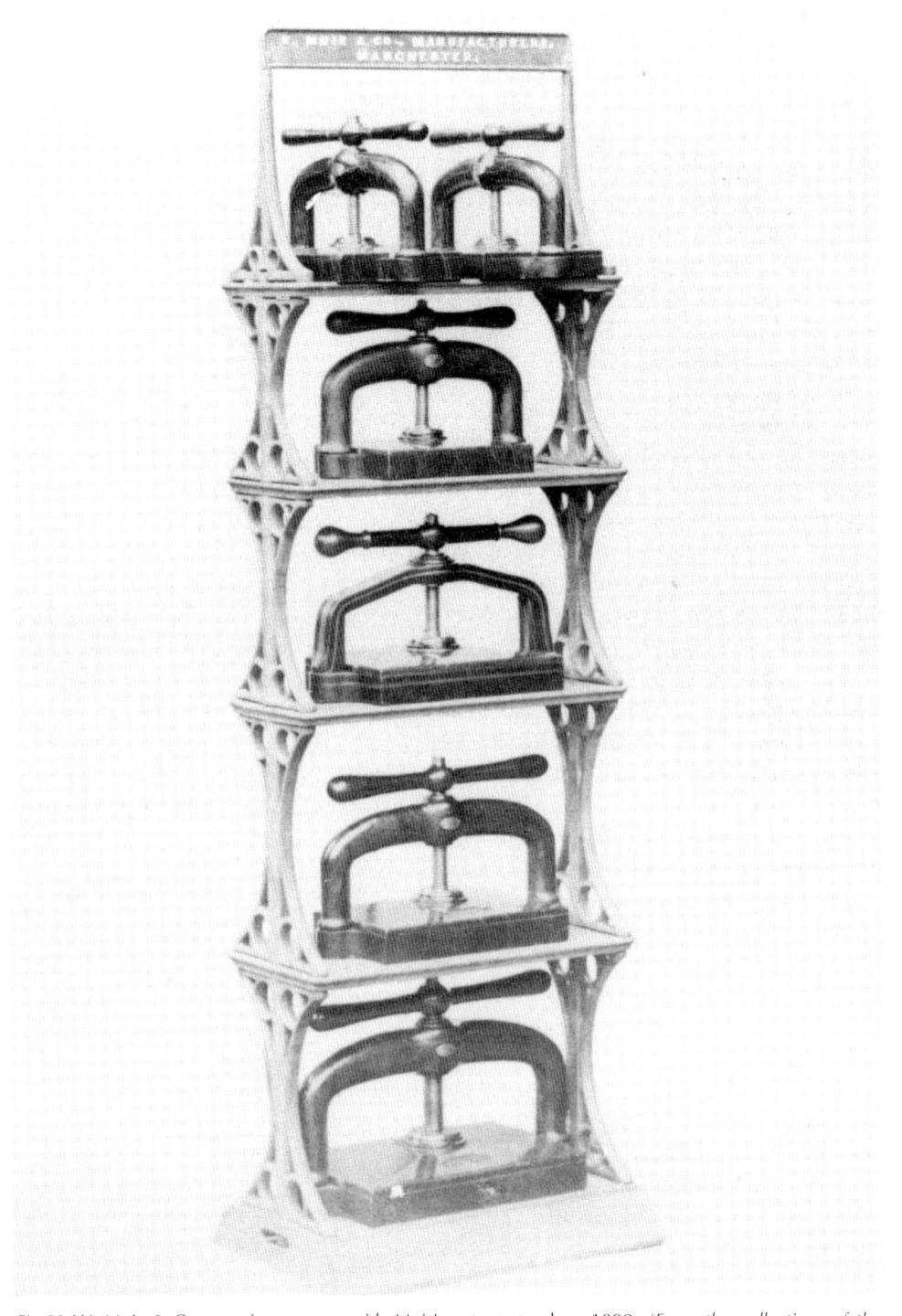

Fig.82 W. Muir & Co. copying presses with Muir's patent stands, c.1890. *(From the collections of the Museum of Science and Industry in Manchester)*

Fig.83 Muir's patent double grindstone apparatus, c.1895. *(Catalogue of W. Muir & Co.)*

copying and embossing presses as well as being machinists and tool makers. Production of the presses probably came to an end in the late 1880s or early 1890s.

The double grindstone was perhaps Muir's strongest piece of design. It was a relatively simple idea, but proved to be practical and of real use to the engineering industry. In other words, it possessed all the qualities associated with the Maudslay school's products. Two grindstones were mounted on spindles in a trough. The stones were turned by an overhead power source, and could move in the same or opposite directions. The stones were in constant contact as they turned, so that they became self-dressing; inequalities on the periphery of each stone were removed by the action of the other, so the grinding surfaces were kept true and even. One stone moved horizontally, with a very slow motion, in order to maintain the continuous contact between them, and their positions could be adjusted as they wore down, by means of worm gear. Water cans mounted above the stones washed the dust down into the trough. The self-dressing action produced much better surfaces than standard grindstones, enabling workmen to grind their edge tools with more accuracy.

Muir patented the grindstone in 1853. An illustration of it was featured in the firm's advertisement in the 1854-55 *Manchester Mercantile & Manufacturing Annual Directory*, and it was exhibited in Paris in 1855, where it won a medal. Muir also received an award for it from the Society of Arts in the same year, after it was shown at their Exhibition of Inventions. The grindstone was still in production in 1895, and the basic design had barely changed over forty years since its development. Models were available with stones ranging from 2ft 6in to 4ft in diameter and from 6 to 12in in thickness. The catalogue proudly said that 'the length of time these Machines have kept in public estimation, and the large number sold, are a sufficient guarantee of their excellence and efficiency, and from the many years we have had them in use, we are confident that the Stones wear much longer than those on the old principle'.

Soon after the move to Britannia Works, Muir entered the field of small-arms and ordnance-making machinery. His first contract was in 1852, for bespoke machine tools for the Royal Arsenal at Woolwich. He then received an order for tools and machinery for manufacturing rifle sights at the new Royal Small Arms Factory at Enfield. The parts for the sights were intended to be interchangeable. Robert Smiles quoted C.F. Partington, a Scientific Demonstrator: 'The rare ingenuity [Muir & Co.] have exhibited has brought us to what may be considered the *magnum bonum* of engineering skill in rifle sight machinery'.[27] Both the Woolwich Arsenal and the Royal Small Arms Factories were awarded medals at the 1862 exhibition for the quality of their workmanship, quality which was in some degree due to Muir's machinery. Muir also made sights for the Minié rifle at Britannia Works, on machinery that was probably very similar to that he supplied to Enfield. He took the contract at the start of the Crimean War, and continued making the sights on a reduced scale after the end of the war in 1856. In August of that year *The Engineer* described the tools Muir was using and noted that the slots in the upright part of the sights were cut 'by a very pretty adaptation of Nasmyth's recessing drill'. The article described the sight workshops as being 'fitted up in the most complete and systematic style to be met with in any establishment in Manchester'. This was not quite an automated production line, however, as the author noted that several men were employed to finish the pieces of the sights and adjust any inaccuracies that were present after the components had left the machines. The firm also supplied machine tools to the Asunción Arsenal in Paraguay. By 1884 the style of 'small arms and ordanance machine makers' had been added to the firm's entries in the Manchester directories, and the 1895 catalogue pointed out that the firm made specialist machines for gun makers. The firm's machinery became very sophisticated. A set of photographs probably dating from around 1890 survives, showing fifteen machines for making rifle stocks and barrel bedding alone.

Muir patented or developed various other machines. These included a machine for cutting sugar loaves into cubes, which employed an arrangement of circular saws and which then sorted the cubes. According to Prosser this was widely employed by the sugar industry, but only long after Muir patented it. Muir also developed machinery for winding cotton balls and bobbins, which Smiles claimed had come into universal use. In 1853 he patented templates for cutting out garments. Two metal templates sandwiched several sheets of fabric and guided the shears or cutters around them. Prosser dismissed Muir's patents and inventions as being of little importance. They may appear so today when compared to the revolutionary developments made by other members of the Maudslay school, but contemporary manufacturers were certainly appreciative of his designs.

Muir left Manchester and moved to London in late 1866 or early 1867, although he remained the nominal head of William Muir & Co. for several years afterwards. He settled in

Brockley, where Robert Smiles noted that he set up a workshop in his garden which included his favourite old foot lathe. He did not stop inventing, and although none of his later patents were particularly practical they show that his interests were still wide-ranging. He became interested in motive power. In 1867 he patented apparatus for propelling trains or carriages which employed chains, wires or ropes laid alongside the tracks, while in 1882 he patented a method of motive power for machinery which used the buoyant effect of air-tight boxes 'arranged in an endless chain-like form' as they rose up through a tank of water. In 1886 he patented an improved method of reciprocating motion for machine tools that required to and fro or vertical travel, such as planing, slotting or shaping machines.

The firm he founded did not forget him. On 31 December 1883 twenty-two of the workforce at Britannia Works sent an address to Muir tendering 'sincere and heartfelt thanks to you as the founder of Messrs Wm. Muir & Co.'[28] Muir died in Brockley on 15 June 1888, after a month's illness. Unlike some of his Maudslay school peers he had enjoyed a long and peaceful retirement. He was buried in Brockley Cemetery on 19 June 1888.

Muir had married Eliza Wellbank Dickinson of Drypool, Yorkshire, in London in August 1832, while he was working for Maudslay's. They had five children, all sons. Eliza died in 1882. Only William, the second son, who died in 1879, was not an engineer. He was a linguist, and travelled extensively in Europe. John Wellbank Muir, their third son, went to Paraguay as an engineer to the Ascunción Arsenal and was killed in a skirmish with Argentinean forces in 1863. Their youngest son Edwin became a civil engineer. He was based in Aberdeen when he became a shareholder in William Muir & Co. Ltd in 1867, but by the time of his father's death had moved to Manchester. The two sons who were most closely connected with the firm and machine tools were the eldest, Andrew, and their fourth son, Alfred. Andrew worked on the machine tools for the Woolwich Arsenal and the Royal Small Arms Factory, Enfield, and by the time of his father's death was running the firm's London office.

However it was Alfred who most resembled his father in his interests and talents. Alfred was born in 1840, and when he had finished school in Manchester he joined his father at Britannia Works. After a period as a journeyman at Penn's Marine Engineering Works in Greenwich he returned to Manchester and took over William Muir & Co. when his father moved to London. Like his father, Alfred insisted on high quality products, and he was equally if not more inventive. Among his many patents were couplings for securing and releasing mandrels and cutters, and capstan slide-rests for lathes. His milling and profiling machines, and the cutters and reamers for milling machines that he developed and patented in 1880, marked another new direction for the firm. They were so successful that on a 1915 advertising sheet the firm was able to boast that 'Amongst Engineers our claim as pioneers of milling and the manufacture of milling machines and milling cutters is generally recognised, as it is now over 40 years since we practically developed this method of machining in England...'. Alfred died on 10 May 1902. Remembering him to the Manchester Association of Engineers in November 1905, Thomas Ashbury said 'He took great pride in turning out work, only the best of its kind, both in finish and accuracy'. Such words could easily have been said about his father.

When William left Britannia Works, Alfred became head of the firm, while Andrew ran the London office. In 1878 Alfred was in partnership with Robert and Charles Garnett. The firm was once again floated as a limited concern in 1893, with Alfred as the Managing Director and the shares mainly divided up between various members of the Garnett and Muir families.[29] When Alfred died in 1902 Herbert Garnett took over as Managing Director, and the influence of the Muir family appears to have waned. Later shareholders included members of the Kearns family and further Garnetts. The firm continued until November 1932 when a receiver was appointed. The assets were sold off by October the next year, and on 13 September 1934 the receiver reported to the Registrar of Companies that the firm had no assets and was carrying on no trading. William Muir & Co. Ltd of Britannia Works was formally dissolved on 4 October 1935.

According to Robert Smiles, William Muir's religion was as close to his heart as engineering. He was a member of the Scottish Presbyterian Church, in which he remained active all his life. When he moved to London to join Maudslay's, he joined the congregation of the Scots Church in Crown Court, Covent Garden. In particular Muir believed passionately in the strict observance of the Sabbath. Robert Smiles describes how Muir and Whitworth differed over this both in London and in Manchester. He suggests that Whitworth's displeasure over Muir's refusal to work on the knitting machine on a Sunday in order to prepare it to go to London was responsible for Muir wanting to leave Chorlton Street. Whether or not this story is true, Muir made his views on the sanctity of the Sabbath plain in his reply to the workers' address on 10 February 1884: 'I was never asked to work on Sundays before I came

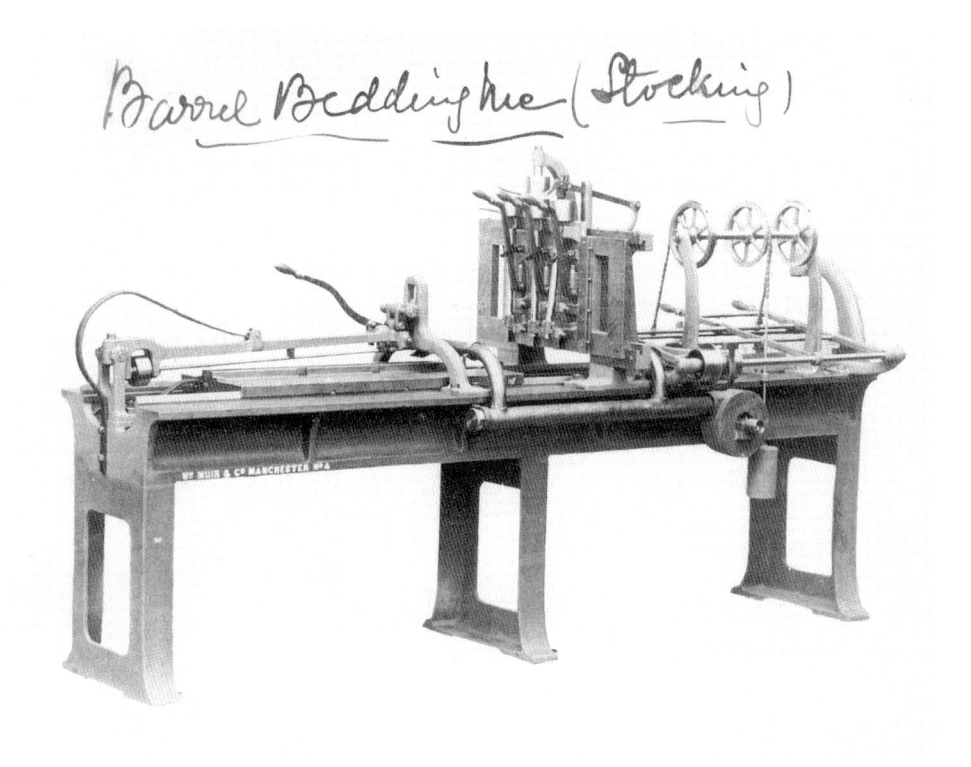

Fig.84 Muir's machine for shaping barrel bedding, c.1890. *(From the collections of the Museum of Science and Industry in Manchester)*

to England, and I am thankful that I have been able to withstand pressure and to resist temptation in this matter. I have ever felt it my duty to do all I could to preserve inviolate the palladium of the working man's birthright in the Day of Rest'. Once again he pointed to his former master Thomas Morton as setting him a shining example.

Muir's deeply held religious beliefs were linked to a concern for the lot of the working man. He strongly sympathised with the Temperance Movement, and he designed frontages for public houses which featured narrow doorways and passages to prevent women with large crinolines entering, and which featured large plate glass windows which rendered the entire interior visible from the outside. He hoped such visibility would reduce drunkenness, and he also proposed that the back rooms of public houses be completely cut off from the bar to prevent people being hidden there. However his designs were not purely prescriptive, as he incorporated heated pavements and drinking fountains into the frontages to give warmth to the destitute poor at night and clean filtered water to the public. Muir patented these arrangements in 1865. He also drew up floor plans for joint coffee and public houses whereby the section where alcohol was served could be completely cut off from the part serving food and coffee by means of metal screens. These arrangements were patented in 1885, and were intended to allow 'working men to have a proper place to go into to have meals without having spirits and ales put before them'. Muir was also one of the few employers in Manchester to keep his works open during the 'Great Lock Out' of 1852, and the arrangement of the firm as a limited liability concern in 1867 may have been an experiment in co-operative ownership with the aim of vesting the workers with an interest in the company.

William Muir was a modest man, and the commentators on his life attributed his lack of fame to this. Robert Smiles called him 'quiet, modest and unassuming almost to a fault' and said that his simple funeral in 1888 reflected his 'unpretentious life'.[30] He took great pride in his work and his firm, and he was not a crusader or a seeker of personal renown. His obituary in *Engineering* perhaps summed him up best:

> *If Mr. Muir did not occupy so important a place in public attention as did some of his contemporaries, it by no means follows that he was behind them in the knowledge of his art; his modest disposition rendered him incapable of pushing himself forward, while his attention was so fully concentrated upon his work that he had no time or thought to bestow upon gaining a widespread personal popularity. His tools carried his name all over the world, and wherever they went they testified to his integrity and ability.*[31]

Endnotes

[1] The main sources of information on Muir's life are Robert Smiles 1888, extracts from which were published in *The Engineer*, 24 August 1888; obituaries in *Engineering*, XLVI (1888), p.194, and *PIME* (1888), pp.440-42; entry in the *DNB* (1894) by Richard Bissell Prosser, drawing on Smiles. Muir's papers and business records are presumed lost, so that some information given in the secondary accounts is unverifiable.

[2] Smiles 1863, p.233.

[3] Armytage 1961, p.128.

[4] Floud 1976, p.183.

[5] Saul 1968, p.23.
[6] R. Smiles 1888, p.6.
[7] Quoted by R. Smiles 1888, p.34.
[8] R. Smiles 1888, pp.18-19. Neither Smiles nor Prosser says that he had charge of the entire works.
[9] R. Smiles 1888, pp.18-20. Smiles' anecdote about Muir knowing Whitworth and Nasmyth at Maudslay's and Whitworth cultivating his acquaintance has been frequently repeated. Muir certainly met Nasmyth, but if the chronology suggested by Norman Atkinson for Whitworth's time in London is correct (Atkinson 1997, p.24), then the two men met while Whitworth was at Joseph Clement's.
[10] For example contrast Roberts or Nasmyth. Whitworth's career most resembles Muir's – he worked for Holtzapffel, L.W. Wright and Joseph Clement after leaving Maudslay, but Atkinson argues that this was a deliberately planned career path and that Whitworth always intended to found his own business: Atkinson 1997, pp.18-19.
[11] For descriptions of Muir's time at Whitworth's and their differences, see R. Smiles 1888, pp.22-25, and Atkinson 1997, pp.134-37.
[12] R. Smiles 1888, p.26.
[13] Holtzapffel 1846, p.505.
[14] R. Smiles 1888, p.28.
[15] CA, Boulton & Watt Archives, Business Letter Book, Vol. 63. The sale was, however, of coining presses and medals from the former Soho Mint.
[16] Institution of Mechanical Engineers Archives, membership application, 8 May 1863.
[17] Atkinson 1997, p.316.
[18] R. Smiles 1888, p.29. The wording, however, is somewhat ambiguous: '[Muir] next, on his own account, leased a large plot of land in Strangeways, Manchester, from Earl Ducie; designed and commenced the erection of the large establishment ... long known as the Britannia Works. After having created and established the business he took partners'. This does not preclude Muir having taken partners before the completion of the works.
[19] Cantrell 1984, pp.15-16.
[20] Saul 1968, p.23.
[21] For the 1867 Memorandum & Articles of Association, correspondence and Dissolution Certificate of 1883, see PRO, BT 31/1317/3411.
[22] 'Well designed and remarkably well executed…Muir's tools are very ingenious, well designed and well made. Some of these remarkable designs are entirely unique to him. Machine-tools from his works are very highly regarded in England': *Exposition Universelle* 1855, pp.287-88.
[23] *Reports by the Juries* 1863, Class VII, p.3.
[24] MSIM, Beyer Peacock & Co. Archives, BCA/1, Company Assets 1854 onwards.
[25] BCA, Boulton & Watt Archives, Business Letter Books, Vols. 135-39.
[26] Saul 1968, p.24.
[27] R. Smiles 1888, p.30.
[28] The address and Muir's reply are reprinted in R. Smiles 1888, pp.40-43 and pp.33-36 respectively.
[29] For documents relating to William Muir & Co. Ltd from 1893 to 1935, see PRO, BT 31/31786/39715.
[30] R. Smiles 1888, pp.7, 39.
[31] *Engineering* 1888, p.194.

10
Maudslay, Sons & Field, 1831-1904

Laurence Ince

By the time of Maudslay's death in February 1831, the foundations were laid of a firm which would flourish until the last decade of the nineteenth century.[1] Maudslay, Sons & Field continued to achieve engineering distinction throughout the century, especially in its marine products, but also through the accomplishments of some notable individuals who had worked or trained at Lambeth.

There had been signs of change in the business before Maudslay died, in particular the growth in importance of marine engineering during the 1820s, and the development of the works to accommodate this. The firm was well placed in London to influence decisions made by important purchasers of marine steam engines, including the Post Office, the Royal Navy, and the Thames steam boat companies which sprang up during the 1820s and 1830s. Maudslay's was in direct competition with several other early builders of marine steam engines such as the Butterley Co., Boulton & Watt, and Fawcett of Liverpool. However, Maudslay's stole an advantage on these competitors by enlarging their Lambeth works to build larger marine steam engines, and bringing a superior style to their products. The firm's new erecting shop was admired by James Watt junior in 1823:

> ...the Freelings have made a visit to Maudslay's premises and speak much of the new building he has to try and shew his engines in, as well as the workmanship and appearance of the engines he has completed for the Navy. Wonder we cannot adopt the same beautiful Gothic in our framing and the same neatness of finish? They found Golbourn who you know has long been a supporter and advocate of Maudslay and has been suspected of being concerned in his works, as well as his highness of Clarence. Somebody must have found money in abundance, as the new buildings are stated to be of great magnitude and magnificence ...[2]

This new erecting shop was 148ft by 55ft with the walls rising to 20ft in height. Its roof was a novel construction of iron plates supported by cast iron arches.

Continuity with Maudslay's engineering ideals was provided by the presence of Joshua Field (1786-1863) as a partner in the business. The other partners of Maudslay, Sons & Field were Thomas Henry Maudslay, John Maudslay and Joseph Maudslay. Field was born in Hackney, the son of a corn merchant. After schooling at Harlow, Essex, he was apprenticed at sixteen as an engineer at the naval dockyard at Portsmouth. There he came under the tutelage of Sir Samuel Bentham, Inspector General of Naval Works. It was Bentham who recommended Field to Henry Maudslay when he was looking for a draughtsman to help him in the planning and construction of the famous block making machines for Portsmouth Dockyard. Field's work on this project must have impressed Maudslay for at some time between 1813 and 1820

Fig.85 Joshua Field (1786-1863).
(Institution of Civil Engineers)

he became a partner in the business. During 1821 Field gathered valuable evidence of the progress of British engineering by visiting the most important works in the Midlands and Lancashire and recording his impressions in a diary. Field was also a founder of the Institution of Civil Engineers and became president in 1848. He was admitted as a Fellow of the Royal Society in 1836.[3] Field continued to play an active part in the management of the firm until his own death in 1863, which came as a great blow to the business. He was considered to be one of the leading engineers of his age and during his time at Maudslay's had taken out a number of patents, some jointly with Henry and Joseph Maudslay.

During the 1830s the business developed its range of products, which included mills, stationary steam engines and engines for ships. Maudslay's also produced machinery for coining. The firm's interest in these products was prompted by Boulton & Watt's sub-contracting parts of their large mint orders. Boulton & Watt would build the engines and coining machines, the Rennies the rolling mill machinery, while Maudslay's produced equipment for the melting and refining departments and the smiths' shops and forges. In 1824 Boulton & Watt completed a mint for Bombay at a cost of £40,000 with Maudslay's part of the order assessed at £4,153.[4] In 1835 Maudslay's built an engine and boiler for one of Dance's steam carriages and in 1838 six locomotives were completed for the London & Birmingham Railway.[5]

The marine steam engine, though, was becoming the staple product of the firm. Its growing importance seems to have been evident to Henry Maudslay at an early juncture for

Fig.86 Joseph Maudslay (1801-1861). *(Science Museum/SSPL)*

he had the foresight to have Joseph, his fourth son, trained as a shipbuilder at the yard of William Pitcher at Northfleet. After his training, Joseph Maudslay (1801-1861) must have joined the firm at the time when some of the early Maudslay marine engines were built at Lambeth. These were side lever engines which were mainly supplied in pairs to drive paddle wheels. The development of the marine engine by the firm lay mainly in the hands of Joseph Maudslay. Between 1827 and 1858 he took out twenty-two patents which mainly related to improvements in marine engine design. The most important of these were for the oscillating marine engine, the 'Siamese' engine with twin vertical cylinders, and the annular engine with double piston rods. Joseph Maudslay was also a pioneer in the field of designs for propellers and in the production of screw engines. He was a founding member of the Institution of Naval Architects and in 1860 contributed a paper to the first volume of the society's transactions.[6]

An additional product was added to Maudslay's list in the 1830s when four complete iron tugs were built for the East India Co. These tugs were to be 125ft in length and 22ft in beam, with each ship powered by two 60hp oscillating engines. One great feat of engineering associated with these ships was that each vessel had a draught of only 22½in. The first of these tugs was completed in 1832 and then tested on the Thames. The ships, after completion, were disassembled and sent out to India on board another larger ship. The first tugs to be completed were *Lord William Bentinck* and *Thames* which were packed into crates and sent out on the East Indiaman, *Larkins*. The tugs *Megna* and *Jumna* followed later, and by 1836 all the tugs were in service pulling 'flats' up and down the Hooghly.[7]

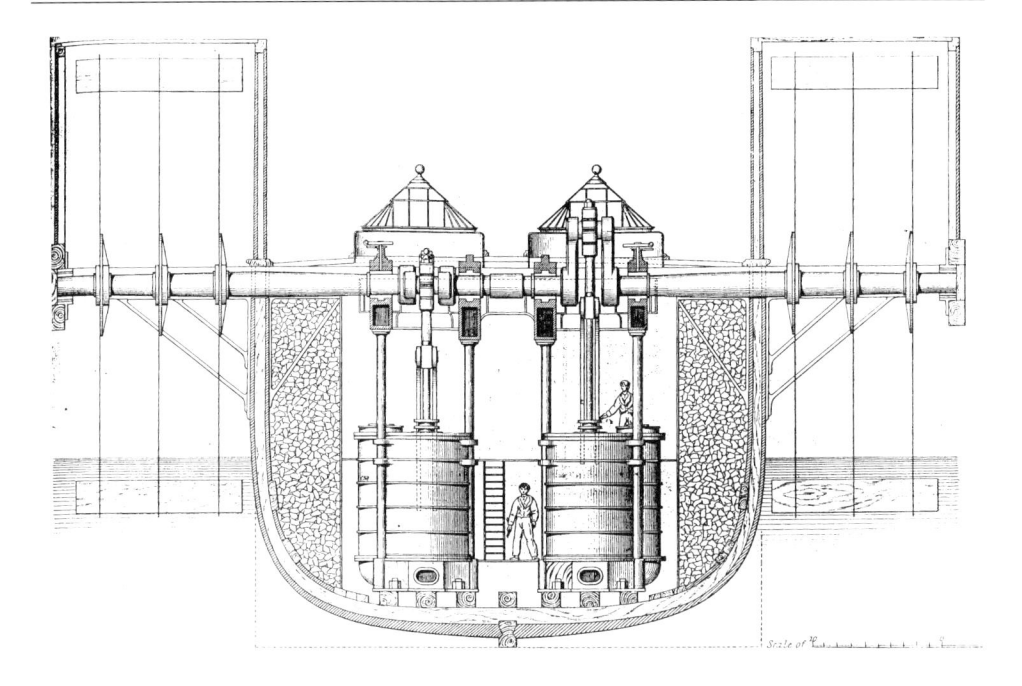

Fig.87 The Maudslay paddle engines for *HMS Retribution*, 1842. This ship was powered by two engines with twin cylinders. Units of this design were known as Siamese engines. *(Bourne, 1861)*

During the 1840s the firm received significant business associated with the boom in railway building in Britain. These were not orders for locomotives but for stationary steam engines. In 1840 the London and Blackwall rope railway, a double line three-and-a-half miles long, was completed under the supervision of Stephenson and Bidder. It ran along a viaduct of brick arches with a falling gradient towards Blackwall. Powerful stationary engines at each end of the line, driving huge winding drums, hauled the trains at twenty-five miles an hour. The two pairs of engines, each rated at 400hp, at the London terminus in Fenchurch Street were by Maudslay, Sons & Field. The engines at the Blackwall terminus were about 280hp, by Barnes of London.[8] Earlier, in 1838, Maudslay's had produced engines for the Euston-Camden incline of the London & Birmingham Railway. The firm was also encouraged to tender for engines for several atmospheric railway lines projected in the 1840s. In around 1846 Maudslay's built engines for Brunel's South Devon Railway, for which they received £11,124, a small proportion of the total expended by this scheme on engines. Boulton & Watt's part of the contract came to £75,319.[9] As this project collapsed the partners at the Lambeth works must have been relieved not to have been further involved.

By the late 1840s Maudslay's could offer a range of stationary steam engines: 200hp beam engines; 110hp beam engines with 55½in cylinders; 60hp beam engines with 43in cylinders; 30hp beam engines with 27in cylinders; 110hp compound beam engines with two 40inch cylinders; 210hp compound beam engines with two 54in cylinders; 400hp compound beam engines with two 72in cylinders; 30hp horizontal engines with 27in cylinders; and 125hp horizontal engines with 55in cylinders.[10]

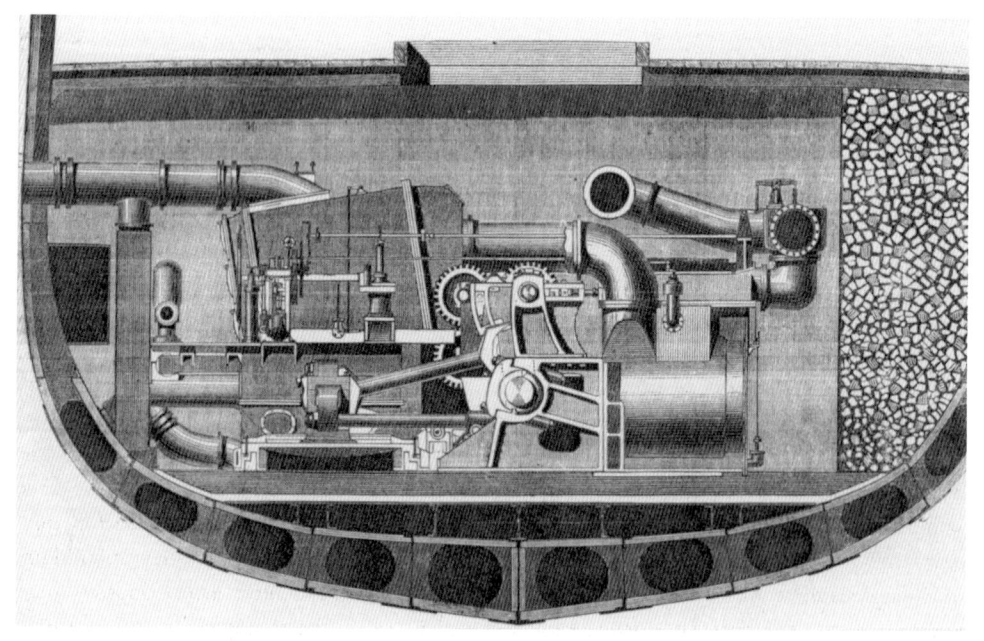

Fig.88 The horizontal double piston rod engine fitted by Maudslay's into the *Jumna*, a ship built by Palmers of Jarrow in 1867. *(The Engineer)*

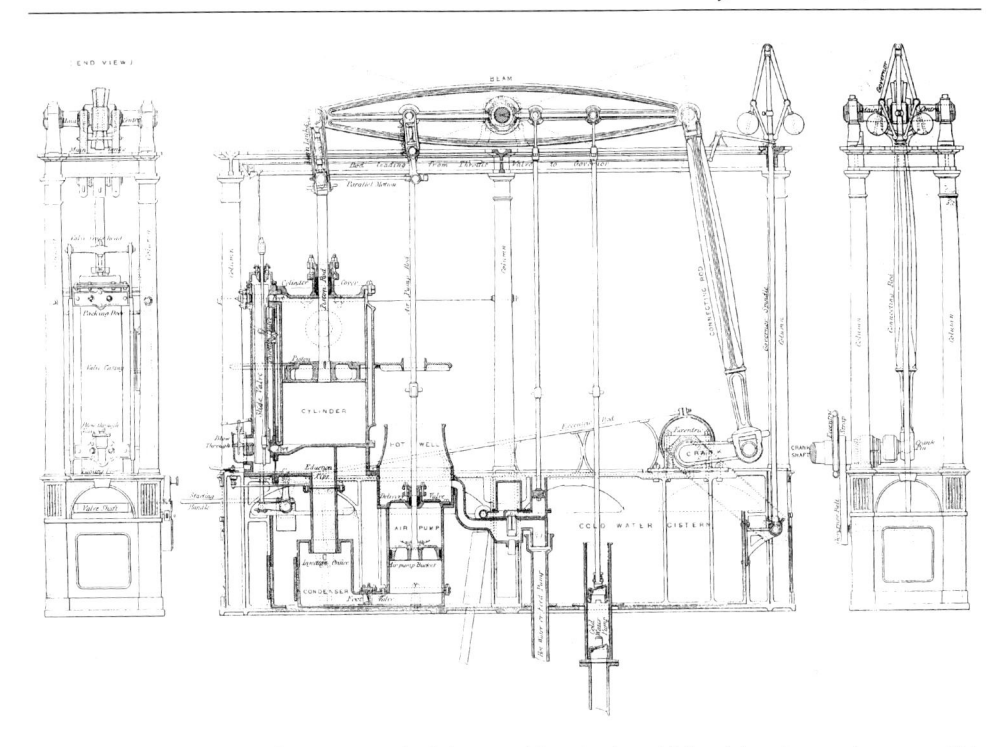

Fig.89 A 16hp six-column beam engine built by Maudslay's in the middle of the nineteenth century. This type of engine would be purchased as a power source for small industrial mills and factories. *(Bourne, 1861)*

Between 1840 and 1853 Maudslay's built stationary steam engines for the Thames Tunnel (30hp); Buxton Waterworks (16hp high pressure); Lisbon Snuff Manufactory (25hp); Sardinia (12hp); Croydon & Epsom Atmospheric Railway (50hp); the Nile (25hp); and Sebastopol Dock pumping engines (30hp).[11] While offering to produce compact engines to the horizontal design the firm also continued to build engines to Maudslay's table design. Two large table engines, supplied to the Constantinople Mint in 1841 and 1853, still survive in their original building. However during the 1840s by far the most important product constructed by Maudslay's was the marine steam engine. In May 1851 it was recorded that the firm had fitted 103 vessels with side lever engines, twenty-eight with oscillating engines, fifty-six with double cylinder engines, twenty-eight with annular engines, six with steeple engines and nineteen with direct-acting screw engines, with a total of twenty-five vessels being powered by screws.[12] This important trade grew because of the expansion of steam ship-building for the navy. The preferred suppliers of marine engines for the Admiralty became the two London firms of Maudslay, Sons & Field, and John Penn. As well as building engines for large naval ships, Maudslay's and several other firms were asked to help the Royal Navy in building small engines for gunboats during the Crimean War. The war saw little enemy opposition at sea, with naval operations confined to blockade and attack on the enemy coast. It soon came apparent that there was a shortage of British naval vessels with the necessary armament and shallow draught for use in the war. A host of builders were encouraged to build a large

Fig.90, above and opposite: Details of a large table engine supplied to the Constantinople Mint by Maudslay's in 1841. *(Courtesy Ian M. Clark)*

number of gunboats and engine makers were given contracts to supply the machinery. Maudslay's during the period 1854-56 produced no fewer than seventy-eight engines for British gunboats and dispatch vessels. Mainly these boats were provided with a high pressure 15½in by 1ft 6in engine making 133 stokes per minute. Each engine was served with three boilers, each 15ft 5in long.[13] The nominal horse-power of these engines was 60 but they produced an indicated power of well over 200. This large contract necessitated the firm expanding their workforce and during the Crimean War 1,200 men were employed at the Lambeth works.[14]

The 1860s were a time of considerable changes for the firm. The partners in Maudslay, Sons & Field at the beginning of 1860 were Thomas Henry Maudslay, Joshua Field and Joseph Maudslay. In June 1860 a new partnership was constituted for twenty-one years which added the names of Henry Maudslay, Herbert Charles Maudslay, Joshua Field junior, Thomas Henry Maudslay junior, Telford Field and Daniel Fitzpatrick.[15] These names suggest a lengthy continuity from the early days of the firm. However, within a few years the new partnership began to loose important members. Henry Maudslay retired in 1861 and Joseph Maudslay died a few months afterwards. Joshua Field senior died in 1863, followed in 1864 by Thomas Henry Maudslay senior. Although the changes in the partnership during the 1860s must have affected the firm, there were enough long-serving employees to ensure that the production and improvement of Maudslay's marine engines continued, with the company still renowned for the construction of marine engines for navies and merchant fleets. One long-serving employee who gave great service to the marine side of the business

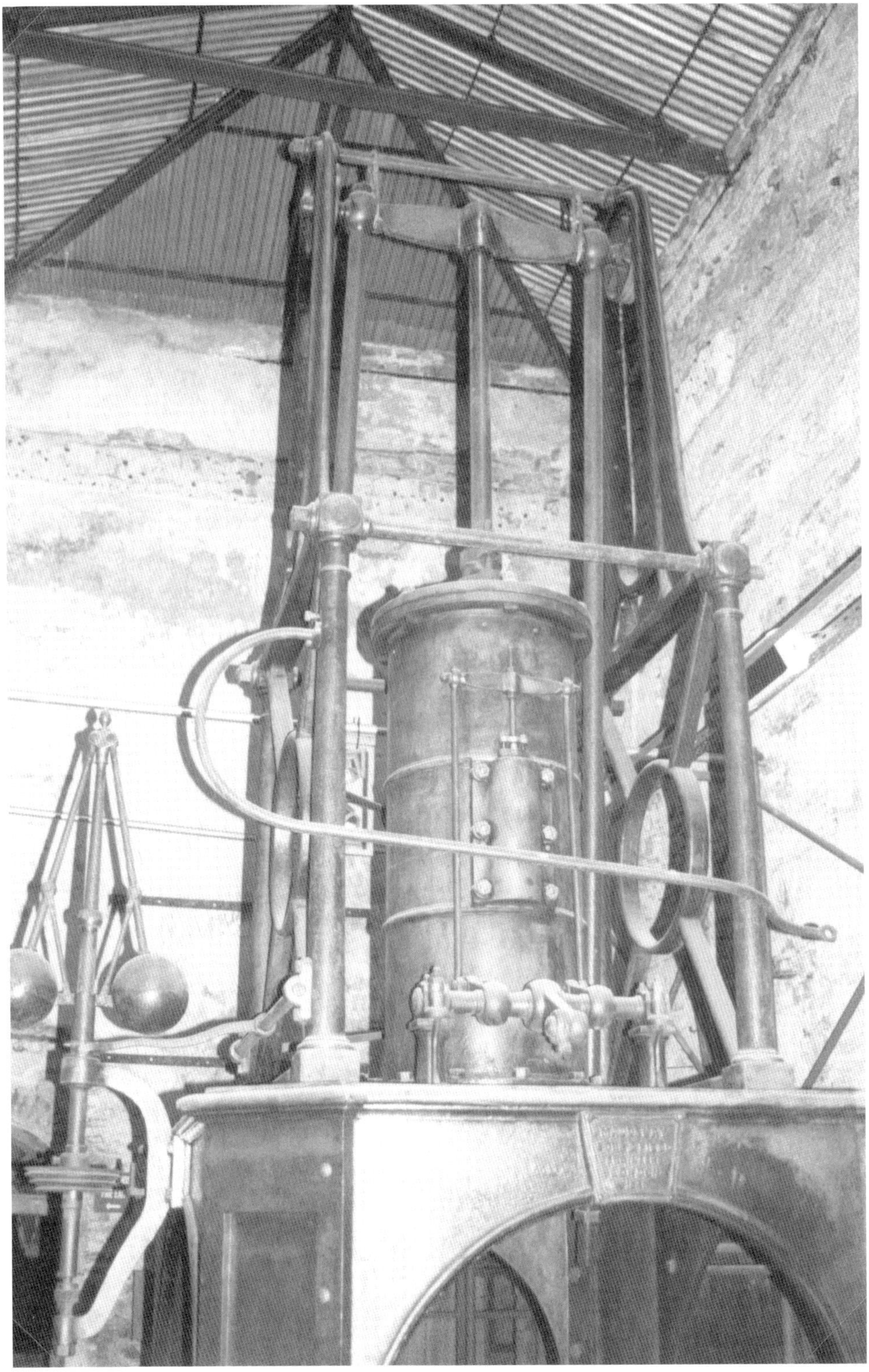

Fig.91 Charles Sells (1820-1900). *(L. Ince)*

was Henry Warriner (1819-1909), a native of Banbury who learnt the engineering trade as an apprentice at Braithwaite & Milner of London. While serving his apprenticeship he became associated with Capt. John Ericsson in fitting the machinery in the pioneer screw steamer *Robert F. Stockton*. He was first employed by the firm in around 1843 to fit out the paddle engines for the *Guiscardo*, a vessel built for the Neapolitan Navy. Warriner later worked for Robert Stephenson and spent a period employed by Ransomes & May, Ipswich. He joined the staff of Maudslay, Sons & Field in 1855 and remained with the firm until the closure of the works, supervising the completion of remaining Admiralty contracts until 1903. He died on 18 March 1909 in his ninetieth year.[16] Charles Sells (1820-1900) was another engineer whose work had an important bearing on the company's success. Sells was chief draughtsman at Lambeth for fifty-eight years from 1847 to 1895, when he retired because of failing eyesight. A Londoner born at Bankside, he went to Maudslay's as a pupil in 1837. When Sells retired the directors of the firm printed a list of vessels which the firm had engined. At the head of the list was quoted a minute of the board announcing his resignation: 'Perhaps no man in England has designed more steam machinery than Mr Sells, commencing from almost the earliest days of the marine engine to the present time … a series of brilliant successes'. They described Sells as 'the most successful designer of the marine steam engine of the present century'.[17]

The 1860s were to be a period of expansion for the firm although the size of the Lambeth works constrained development. As marine engines became larger, it was difficult for the company to satisfy demand for its products. With space at a premium, it seems that the building of stationary engines was primarily affected. The works built only the occasional land engine after the new partnership was formed. Another London site was acquired for some of the overflow work, in 1865 when a yard at East Greenwich was purchased. As well as providing extra space for the firm, this enabled Maudslay's to enter the shipbuilding trade. Between 1865 and 1873 several ships were built by the company at East Greenwich. Two of the largest were the iron sailing clippers *Halloween* and *Blackadder*, completed in 1870. *Blackadder* was dismasted on her maiden voyage, resulting in lawsuits between her owners, underwriters and builders. Both ships proved to be fast and *Halloween* went out to Sydney in sixty-nine days and during one period of twenty-four hours travelled 360 miles. However, the foray into shipbuilding did not prove successful. Maudslay's launched its last ship in 1873 and the yard appears to have then been partially turned over to boiler-making.[18]

The constraints of the Lambeth site caused a continuing and increasing problem for the development of marine engine building. As engine sizes increased, the problem worsened. As early as 1848 the company was undertaking Admiralty contracts for large marine engines. In that year screw engines were built for the steam guard ship HMS *Ajax*. These were horizontal direct acting engines with 55in by 30in cylinders. Soon after, the company began to build horizontal return connecting rod engines. A pair was supplied in 1865 for the iron armour-plated cruiser HMS *Agincourt*. These engines had 101in by 54in cylinders with slide valves actuated by link motion. Maudslay's also adopted the compound design for their marine engines at an early date, for in 1860 the firm completed compound engines for HMS *Octavia*, one of the first naval vessels to receive such engines.

Through many changes in the partnership, members of the Maudslay and Field families continued to dominate. Henry Maudslay retired from the company in 1867 and Henry Charles Maudslay followed in 1868. These vacancies were filled by Walter Henry Maudslay, a younger son of Joseph Maudslay, and the Hon. George Duncan, brother of the third Earl of Camperdown. Again the vitality of the concern was not obviously affected by these changes, with the company building engines for the British, French, Brazilian, Italian, Argentinian, Greek and Chinese navies during the 1870s. Marine engines were also supplied to the British, Russian and French merchant fleets. In 1870 a report on engines for HMS *Druid* pointed out that, as they were Maudslay products, 'therefore it is unnecessary to comment on their exquisite finish'.[19] The firm could still be enticed to build stationary engines for prestigious and noteworthy undertakings. Some of its last such engines, completed in about 1869 for the Indus Tunnel at Attock, were two Cornish beam pumping engines with 36in by 8ft cylinders. Each engine beam was 24ft long and made up of two plates. Maudslay's also built the boiler house roof for these engines at East Greenwich, of a wrought iron construction covered with heavy galvanised corrugated plates.[20] Another prestigious stationary engine was completed in the early 1880s, designed to generate electricity for the Royal Mint.[21] It was a compound tandem engine which had the look of a marine engine, an indication of where the real market was for Maudslay's engineering expertise.

Problems with the old and cramped premises at Lambeth took their toll on production efficiency. In 1881 a visitor recorded that refurbishment was in progress as previous piecemeal extensions had made the flow of work through different departments very inconvenient. The foundry, 'in which are produced some of the heaviest castings ever made for marine engines'

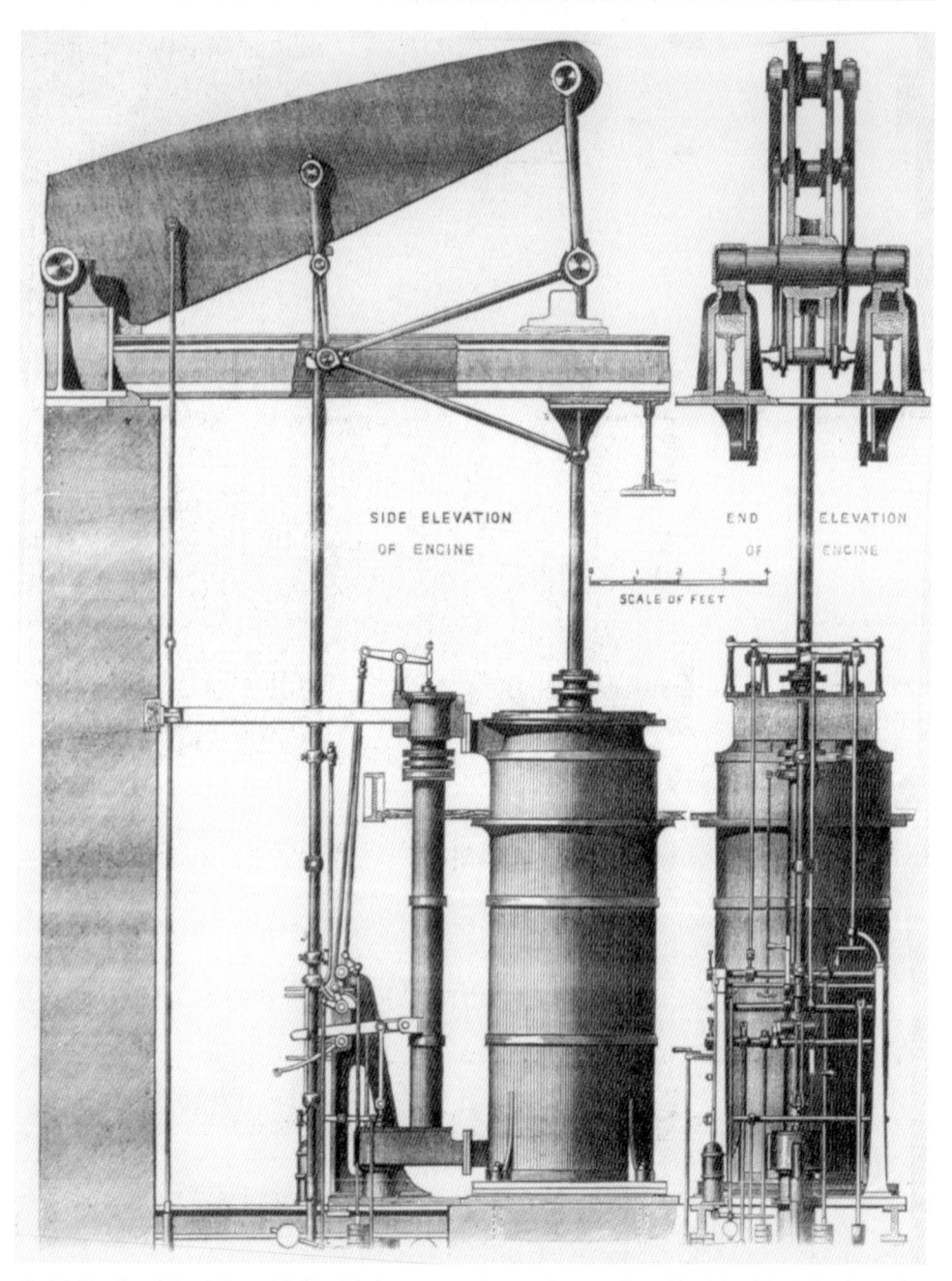

Fig.92 Details of the 36in x 8ft Cornish beam pumping engines produced by Maudslay's for the Indus Tunnel, Attock, India. This pair of engines was completed in the late 1860s. *(The Engineer)*

Fig.93 Walter Henry Maudslay (1844-1927). *(L. Ince)*

was approached by a flight of stairs as it had to be above ground level to avoid the flooding of deep castings. Yet the factory had 'certain of the finest tools ever constructed in the turnery and machine shops', some of them by Smith, Beacock & Tannett of Leeds. Maudslays had been involved in a long-running chancery case relating to the Lambeth premises, and once this was settled had decided to move their entire boiler-making operation to East Greenwich. As a result, reported the visitor, they were vacating the 'front erecting shop', the lease of which had expired. This shop had 'one of the first iron roofs if not the first', a hipped construction with cast iron principals erected by Henry Maudslay himself. In the main erecting shop, which had a similar roof, the compound engine of the *Colossus* was being built. The old boiler shop was under conversion to another erecting shop. Because of the alterations in progress, few engines or boilers were being finished, although the visitor noted engines for the *Triton*, a surveying ship, and large oscillating engines for the continental boats of the London, Chatham & Dover Railway Co.[22]

The strategy of moving the complete boiler-making department to East Greenwich seems to have been a success, for the works continued engine building for another twenty years. The continued survival of the company can be attributed to its last managing director, Walter Henry Maudslay. He was born in 1844, the third son of Henry Maudslay's son, Joseph. W.H. Maudslay was educated at Brighton College and Trinity College, Cambridge, before joining the firm. His duties were commercial and administrative rather than technical, in which fields he showed much flair. In 1893 as managing director he secured the British rights in the Belleville water-tube boiler and in 1903 on the liquidation of the business established the Birmingham Aluminium Casting Co. Other interests lay in industrial insurance and he was a vice-president of the Federated Employers of Great Britain. He died at Falmouth on 25 August 1927.[23]

In order to attract more investment and improve the firm's long-term outlook, W.H. Maudslay oversaw the formation of a limited company in 1889.[24] While this did introduce more capital, many of the shares were still held by the Maudslay and Field families, represented by W.H. Maudslay and Joshua Field. The firm continued to tender for and build large marine steam engines but a rather dangerous precedent was set in 1886 when Maudslay's agreed to build a set of engines for *Sicilia*, an Italian naval ship. These were built in Italy to Maudslay's designs using local labour. Between 1888 and 1904 this pattern was followed on nineteen occasions, as Maudslay engines were built for the navies of Spain, Italy and Argentina. This strategy must have cut down on expenses but the firm lost some control and profit. One wonders if the labour used in these foreign countries also learned skills and knowledge that encouraged local builders to try their hand in marine engine building.

The purchase in 1893 of rights to the Belleville boiler for marine steam engines was a significant one. This improved form of water-tube boiler was invented by Julien Belleville in about 1879 and developed by Messrs Delaunay, Belleville et Cie., of St Denis, France.[25] Each boiler consisted of a set of elements or continuous tubes running to and fro ten times over a furnace, always inclining upwards and finally opening into a steam collector. The Belleville boiler was used in French naval and merchant ships and Royal Navy officers wanted to test it in Britain. In 1892 the gunboat *Sharpshooter* had her locomotive type boilers replaced by eight Belleville boilers working at 245psi. Trials in 1894 showed the new boiler type to be economic, reliable and capable of a high steaming rate.[26] By this time W.H. Maudslay had pounced and sewn up the British rights for the boiler. The admiralty then chose Belleville boilers for the cruisers *Powerful* and *Terrible*. So successful were trials on board these ships that other Belleville installations were approved, water-tube boilers became the standard unit for battleships and cruisers, and it was to Maudslay, Sons & Field that the orders had to come.

During the late 1890s the production of Belleville boilers appears to have been the most important aspect of Maudslay's engineering work. The huge scale of building steam engines for battleships taxed the firm's resources in their cramped buildings. A portfolio of engravings and photographs published by the company in 1892 contains many pages of smaller auxiliary steam engines then needed by naval ships in addition to their main engines. These included air compressing engines designed for furnishing the motive power for Whitehead torpedoes, feed engines, centrifugal pumping engines and engines for generating electricity for lighting.[27] To cater for the range and intensity of business, large-scale management changes were made in 1895. John Sampson, a director since 1892, became departmental manager overseeing foreign business. Algernon Sydney Field was appointed a director of the company in March 1893 on the retirement of his father Joshua. He was also manager of the auxiliary machinery and filters' department. Capt. Allen was made a director in March 1893 and served as a departmental manager.[28] John Sampson (1861-1925) was a Cornish engineer who had served his apprenticeship with the firm. After a period at sea he returned to Lambeth to work for Maudslay's, travelling extensively for the firm in Russia, Italy and Austria and remaining a director until the demise of the company. Sampson was also a director of Thomas Frith & Sons and after the closure of Maudslay's was associated with Harland & Wolff and John Brown & Co.[29]

Although Maudslay's was still receiving orders for marine engines during the 1890s there was a marked reduction in engine building at Lambeth. The change to limited liability had not solved the financial problems and in 1895 the size of the company's mortgage was increased. On 26 June 1896 an overall loss on trading was reported.[30] The directors recognised a serious commercial problem and in January 1897 brought Sir Reginald Gipps on to the board. Large marine engines were paid for in instalments as they were constructed but on 13 January 1897 the directors noted that the instalment payments due to the firm were insufficient to pay wages at the end of the week.[31] While the building of large marine steam engines continued, during the period 1895-1899 only six sets were completed. However a considerable trade had built up in two other sectors of the marine engineering business. The orders for Belleville boilers from other firms building marine engines had reached a very healthy state and also the company had developed a large market for the supply of their auxiliary marine engines and pumps.

During 1898 the company began to build their last set of marine steam engines which had been ordered by the Admiralty for HMS *Venerable*. In the same year Maudslay's were building four main and four auxiliary feed pumping engines for HMS *Euryalus*, under construction by Vickers, Sons & Maxim; feed pumping engines for a battleship of the Formidable class being engined by Earle's Shipbuilding Co. of Hull; and six duplex force and bilge pumps for the Baltic Shipbuilding Works.[32] Later that year a further order was received from Vickers for two drain pumps, four hot well pumps and four combined air and circulating pumps for HMS *Hogue* and HMS *Euryalus*. Maudslay's also appears to have obtained the rights to build Belleville boilers in other countries outside Britain. However the company was so desperate for money at that time that sub-concessions were sold to other engineers. In October 1898 it was noted that the firm had won a contract from the Austro-Hungarian government to build Belleville boilers for two coast defence ships. This

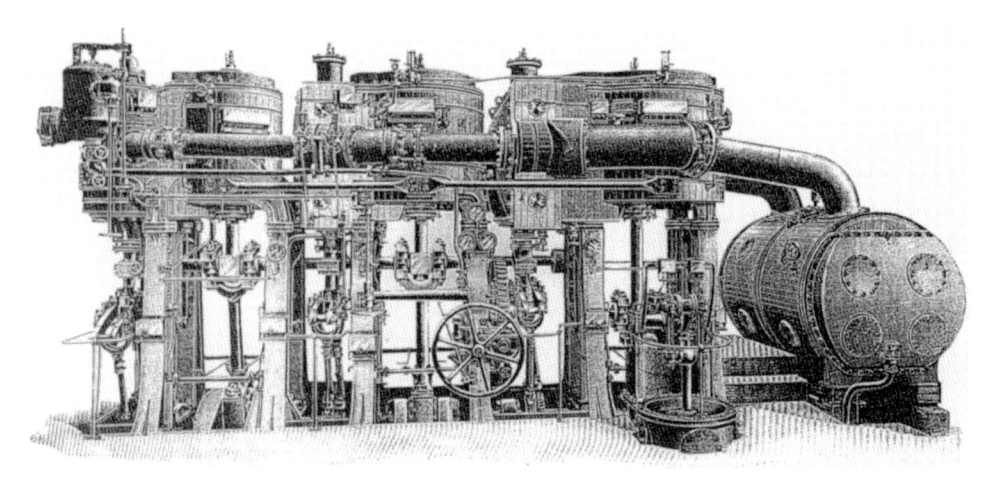

Fig.94 One of a pair of triple expansion engines completed by Maudslay's in 1895 for the 330ft long Russian naval ironclad *Admiral Ushakov*. *(Engineering)*

Fig.95 The Great Wheel at Earl's Court which was built in 1895 by Maudslay's and powered by one of their marine engines. *(L. Ince)*

order was immediately granted as a sub-concession to Durr Gehre & Co. of Austria. The same happened when Italian orders for Belleville boilers were won by Maudslay's. They were given as a sub-concession to Orlando of Livorno and Hawthorn Supply Co. of Naples.[33] Another series of orders came through the initiative of Walter Basset, a naval officer who later joined the firm as a director. Basset's novel idea was to build a Great Wheel at Earl's Court constructed under licence from Lt J.W. Graydon who had been responsible for a smaller wheel at the Chicago Exposition of 1893. The steelwork, by Arrol's Bridge and Roof Co., was assembled at the East Greenwich boiler works. Maudslay's also constructed the marine steam engine which powered the ride. The 270ft diameter wheel, completed in 1895, was an immediate success, with the company building parts for more wheels: Blackpool, 200ft diameter in 1896; Vienna, 200ft, 1897; Paris, 300ft, 1898.[34]

These orders provided the company with only a little breathing space, for the directors realised that great changes had to be made if the firm was to survive. The balance sheet for the year ending 31 December 1897 showed a net loss of £37,205.[35] The board believed that the main problem was the old-fashioned, cramped and inefficient buildings, and began to look for alternative sites. During 1899 the company received no orders for marine steam engines but did gain many contracts to supply auxiliary engines for naval ships. The size of such orders can be gauged from the example recorded by the company on 27 January. It was from Vickers for the first class cruiser HMS *King Alfred* and included twenty double breasted fans and direct acting steam engines, eight furnace air pumping compound engines, four compound fire and bilge pumps, two hotwell engines of the tandem compound type, a steam pumping engine of the compound type for latrine purposes and a drain tank pump.[36] These and other orders gave the board confidence to look for another site to continue operations. By September 1899 the company was trying to purchase industrial buildings at Ipswich but negotiations were swiftly brought to an end when on 11 October three actions by shareholders and debenture-holders led to Maudslay's being put into the hands of receivers.

The contents of the Lambeth works were later put up for auction but some of the machine tools were purchased by Maudslay's for removal to the East Greenwich works.[37] The receivers obviously believed that the company could still operate and make profits from that site. On 22 March 1900 the firm removed some machines to East Greenwich and in July some furniture from Lambeth was also taken to the boiler works. This strategy was adopted by the receivers because they believed that Maudslay's could still exploit the licence to make Belleville boilers and so make money to pay off the accrued debts. However, disaster lay in wait for the newly constituted Maudslay, Sons & Field. The good initial results of the Belleville boiler trials did not continue into service and as time went on more problems and failures began to appear with the operation of the new boilers. A committee of enquiry met under Admiral Sir Compton Domville to investigate the matter. The interim report reported favourably on the adoption of water tube boilers but several unfavourable comments were made respecting Belleville boilers. After these criticisms the journal *Engineering* predicted that '…the Belleville boiler will doubtless cease to appear in future ships of the British Navy'.[38] Without this product the East Greenwich works was forced to survive on boiler repairs and the manufacture of small steam engine parts until final closure in 1904.

Before and after the death of its founder, Maudslay's firm recorded many milestones in marine engineering. In 1816 they supplied a pair of 20in by 2ft 8in engines for *Regent*, the first steamship built on the Thames. Five years later they supplied a pair of 36in by 2½ft engines for the paddle steamer *Rising Star*, built by Daniel Brent of Rotherhithe and intended as a warship in the Chilean revolution, which became the first steamer to cross the Atlantic in a westerly direction.[39] In 1825 the company installed a pair of 43in by 4½ft engines in the *Enterprize*, the first steamship to travel from Britain to India. Complementing this achievement the firm provided two 48¼in by 4½ft engines in 1828 for the Bombay-built ship *Hugh Lindsay* which in 1830 became the first steamer to sail from India to Britain.[40] During the 1820s Maudslays built several sets of engines for some of the first steamships commissioned into the Royal Navy. *Curaçao* built at Dover in 1826 and provided with a pair of 40in by 4ft Maudslay engines became the first steamship in the Netherlands Navy and the first Dutch steamer to cross the Atlantic.[41] The Royal Navy also had an early transatlantic steamer powered by Maudslay engines, *Rhadamanthus*, built in 1832 with two 55½in by 5ft engines.[42] In 1838 the firm engined the *Great Western*, a close second to the *Sirius* in crossing the Atlantic using continuous steam power. Maudslay's also received royal patronage by building paddle engines for two royal yachts, the *Victoria and Albert* (1842) and the *Osborne* (1870).[43] When screw propulsion was first introduced Maudslay's played an important part in the development of this new motive power. In 1841 the firm built two 40in by 4ft engines for HMS *Rattler*, the first screw steamer in the Royal Navy. Twenty years later it was Maudslay's which supplied some of the Royal Navy's earliest compound engines, and in 1876 the engines for the despatch vessels HMSs *Iris* and *Mercury*, the first naval ships to be made of steel.[44]

A further legacy at home and abroad came in the body of engineers who trained or worked at the Lambeth workshops and went on to make notable contributions to the development of marine engineering and other branches of the profession. One such was Richard Sennett, the first naval engineer to hold the office of Engineer-in-Chief to the Royal Navy, who later resigned to join the Maudslay board.[45] By the time he died (at a young age, from consumption) in 1891, Sennett had completed his classic book, *The Marine Steam Engine* which went through many editions and was prescribed reading for any marine engineer.[46] A.C. Kirk, who had been a Maudslay draughtsman, designed what were probably the first triple expansion engines afloat, those of the *Propontis*, built by John Elder & Co.[47] Another Maudslay pupil who achieved fame was Charles Brown (1827-1905), who learnt his profession under Charles Sells at Lambeth and then helped the Swiss firm of Sulzer Brothers gain a European reputation as precision engineers. Henry Sulzer-Steiner said of Brown that he was, 'no less than the founder of the mechanical industry in Switzerland'.[48] Other Maudslay men had connections with the firms or legacies of Maudslay's former protégés: Charles O'Connor (1823-1903), a Maudslay works manager, served part of his apprenticeship with Sharp, Roberts & Co.[49] while Louis Martineau (1866-1895) obtained a Whitworth exhibition after three years in the Maudslay drawing office.[50] William Henry Barlow moved on to be Engineer-in-Charge of the Midland Railway, then a consulting engineer who with Sir John Hawkshaw completed the Clifton Suspension Bridge, and became president of the Institution of Civil Engineers in 1879-80 and vice-president of the Royal Society in 1880.[51] Ebenezer Goddard (1816-82) became a gas lighting engineer,[52] while John Imray (1820-1902) was later President of the Institute of Patent Agents.[53] Less well-known were those who took positions abroad, such as William Blake Lambert (1816-74), who worked for Maudslay's for twelve years and subsequently became Engineer-in-Chief to the Russian fleet,[54] or William Miller (1814-87), who spent the first eight years of his working life at Lambeth and went on to gain honours from the Russian government for his work as superintending engineer for the Black Sea fleet and naval ports, and then chief engineer for a large rocket factory.[55] Nor did the Maudslay family itself rest on its laurels, for after the company's demise R.W. Maudslay, fourth of the engineering dynasty, took the name to the Midlands where he started the Maudslay Motor Co. in 1903 to continue the engineering tradition founded a century earlier by his great-grandfather.

Endnotes

[1] The main sources on Maudslay's business after 1831 are Petree 1949; BCA, Boulton & Watt papers, Matthew Boulton papers; LMA, Maudslay, Sons & Field, draft minute books (A/CC/3639/1-2) and order book, 1889-1904 (A/CC/3639/5); SMAC, notebooks of Charles Sells, 1842-83 (SELLS 1-12) and portfolio of marine engines, boilers and machinery, 1892, Maudslay, Sons & Field Ltd.

[2] BCA, Matthew Boulton Papers, Box 355, to Matthew Boulton, 1 September 1823. Sir Francis Freeling (1764-1836) was for many years secretary to the General Post Office, his son Sir George Henry (1789-1841) commissioner of customs. Henry Goulborn (1784-1856), chief secretary for Ireland was later chancellor of the exchequer.

[3] Petree 1949, pp.24-26.

[4] BCA, Boulton and Watt papers, Box 13, bundle 1, Bombay Mint Costs, 23 March 1824.

[5] Petree 1949, pp.38-39.

[6] Petree 1949, p.26.

[7] Sutton 1981, pp.129-31.

[8] Sandham 1885, p.141.

[9] Hadfield 1985, p.172

[10] Notebooks of Charles Sells: SMAC, SELLS 1, p.49.

[11] Notebooks of Charles Sells: SMAC, SELLS 1, p.49.

[12] Notebooks of Charles Sells: SMAC, SELLS 3, p.177.

[13] Notebooks of Charles Sells: SMAC, SELLS 4, p.90.

[14] Petree 1949, p.22.

[15] Petree 1949, p.22.

[16] Petree 1949, p.28.

[17] Petree 1949, p.28.

[18] Banbury 1971, pp.202-203.

[19] The Engineer, 1 July 1870, pp.6-8.

[20] The Engineer, 8 Apr. 1870, pp.204-206.

[21] Cooper 1988, p.217.

[22] The Engineer, 7 Oct. 1881, p.253.

[23] Petree 1949, p.29.

[24] PRO, BT 31/4410/28681; Draft Minute Book, 1889-1896: LMA, A/CC/3639/1.

[25] Spratt 1951, pp.84-85.

[26] Griffiths 1997, p.127.

[27] Portfolio of Photographs of Marine Engines, Boilers and Machinery, 1892, Maudslay, Sons & Field Ltd.: SMAC, Pc 6609, Ms 225.

[28] Draft Minute Book, Extraordinary General Meeting, 5 April 1895: LMA, A/CC/3639/1.
[29] Petree 1949, p.32.
[30] Draft Minute Book, 26 June 1896: LMA, A/CC/3639/1.
[31] Draft Minute Book, 13 January 1897: LMA, A/CC/3639/2.
[32] Order Book, 1898-1902: LMA, A/CC/3639/5.
[33] Draft Minute Book, 1 October 1898: LMA, A/CC/3639/2.
[34] Petree 1949, p.32.
[35] Draft Minute Book, 9 August 1899: LMA, A/CC/3639/2.
[36] Order Book, 27 January 1899: LMA, A/CC/3639/5.
[37] Order Book, 21 July 1900: LMA, A/CC/3639/5.
[38] *Engineering*, 15 March 1901.
[39] Spratt 1950, pp.8-9.
[40] Banbury 1971, p.198.
[41] Spratt 1950, pp.10-11.
[42] Spratt 1950, p.11.
[43] Petree 1949, p.43.
[44] Petree 1949, p.43.
[45] Petree 1949, p.32.
[46] My copy of this book is dated 1918 and is the thirteenth edition, revised by H.J. Oram.
[47] Smith 1937, p.244.
[48] Smith 1937, p.360.
[49] *PIME* (1903), pp.923-24.
[50] *PICE*, 122 (1894-95), p,399.
[51] *PICE*, 151 (1902-03), pp.388-400.
[52] *PICE*, 72 (1882-83), pp.312-14.
[53] *PIME* (1902), pp.1026-28.
[54] *PICE*, 40 (1874-75), pp.258-59.
[55] *PICE*, 91 (1887-88), p.452.

Bibliography

Allen, T., *The Parish of Lambeth* (1826).
Armytage, W.H.G., *A Social History of Engineering* (1961).
Ashbury, T., *The Jubilee of the Manchester Association of Engineers* (1905).
Ashby, E., *Technology and the Academics* (1959).
Atkinson, N., *Sir Joseph Whitworth* (1996).
Babbage, C., *Passages in the life of a Philosopher* (1864).
Bailey, W.H., 'Richard Roberts the Inventor', *Papers of the Manchester Literary Club*, 5 (1878-79).
Baker, T., *Elements of Mechanism*, second edition, (1858-59).
Banbury, P., *Shipbuilders of the Thames and Medway* (1971).
Barry, P., *Dockyard Economy and Naval Power* (1863).
Bennett, A. R., *The Chronicles of Boulton's Sidings* (1927).
Boucher, C.T.G., *John Rennie 1761-1821. The Life and Work of a Great Engineer* (1963).
Bourne, J., *A Treatise on the Steam Engine* (1861).
Bremner, D., *The Industries of Scotland – Their Rise, Progress, and Present Conditions* (1864).
Brown, P.A.H., *London Publishers and Printers c.1800-1870* (1982).
Brunel, I., *Life of I. K. Brunel* (1870).
Buchanan, R., *Practical Essays on Mill Work*, third edition, revised by G. Rennie (1841).
Buchanan, R.A., 'Trade Unions and Public Opinion 1850-1875', unpublished PhD thesis, University of Cambridge (1957).
Buchanan, R.A., *The Engineers: A History of the Engineering Profession in Britain 1750-1914* (1989).
Buchanan, R.A., *Brunel: The Life and Times of Isambard Kingdom Brunel* (2002).
Burnett, J., (ed.), *Useful Toil: autobiographies of working people from the 1820s to the 1920s* (1974).
Burt, R., *John Taylor, mining entrepreneur and engineer 1779-1863* (1977).
Campbell, R.H., *Carron Company* (1961).
Cantrell, J.A., 'James Nasmyth and the Bridgewater Foundry: Partners and Partnerships', *Business History*, 23 (1981), 346-58.
Cantrell, J.A., *James Nasmyth and the Bridgewater Foundry. A study of entrepreneurship in the early engineering industry* (1984).
Cantrell, J.A., 'James Nasmyth and the Steam Hammer', *TNS*, 56 (1985), 133-38.
Cantrell, J.A., 'The Maudslay Legacy', *Kew Bridge Steam Museum: Maudslay, Sons & Field – the seminar and exhibition* (CD ROM, 2001).
Cantrell, J.A., 'Two Maudslay protégés: the case of Francis Lewis and George Nasmyth', *TNS*, forthcoming (2002).
Cardell, W.P., *Drafts on the memory of a Septuagenarian plain man* (1912).
Carlyle, A. (ed.), *Letters of Jane Carlyle* (1903).
Catling, H., *The Spinning Mule* (1970).
Catterall, G.S., 'The Life and Work of Richard Roberts with special reference to the development of the self actor mule', M.Sc. Thesis, Manchester, UMIST (1975).

Chaloner, W.H., 'New light on Richard Roberts, textile engineer (1789-1864)', *TNS*, 41 (1968-1969).
Chapman, A., 'James Nasmyth: Astronomer of Fire', in P. Moore (ed.), *The Yearbook of Astronomy 1997* (1996), pp.143-67.
Chapman, A., *The Victorian Amateur Astronomer* (1998).
Chrimes, M., Skempton A., Dennison, R.W., Cox, R.C., Ruddock, E. and Cross-Rudkin, P. (eds.), *Biographical Dictionary of Civil Engineers in Great Britain and Ireland – Volume 1: 1500-1830* (2002).
Clark, D.K., *Railway Locomotives* (1860).
Colburn, Z., *Locomotive Engineering, and the Mechanism of Railways* (1871).
Cooksey, J.C.B., *Alexander Nasmyth 1758-1840: a man of the Scottish renaissance* (1991).
Cookson, G., 'The West Yorkshire textile engineering industry, 1780-1850', unpublished D. Phil. Thesis, University of York, 1994.
Cookson, G., 'Millwrights, Clockmakers and the Origins of Textile Machine-Making in Yorkshire', *Textile History*, 27 (1) (1996), 43-57.
Cookson, G., 'Reconstructing a lost engineer: Fleeming Jenkin and problems of sources', in A. Jarvis and K. Smith (eds.), *Perceptions of Great Engineers* (1998).
Cookson, G. & Hempstead, C. A., *A Victorian Scientist And Engineer: Fleeming Jenkin and the Birth of Electrical Engineering* (2000).
Cooper, D.R., *The Art and Craft of Coinmaking* (1988).
Dickinson, H.W., *A Short History of the Steam Engine* (1939).
Dickinson, H.W., 'Joseph Bramah and his Inventions', *TNS*, 22 (1941-42), 169-86.
Dickinson, H.W., 'Richard Roberts, his Life and Inventions', *TNS*, 25 (1945-47), 123-37.
Donkin, S.B., 'Bryan Donkin 1768-1855', *TNS*, 27 (1949-51), 85-94.
Ellison, T., *The Cotton Trade of Great Britain* (1886).
Encyclopaedia Britannica, fourth edition, (1824).
Engels, F. (ed.), with introduction by W.H. Chaloner and W.O. Henderson, *The Condition of the Working Classes in Manchester in 1844* (1958).
English, W., *The Textile Industry* (1969).
Evans, F.T., 'The Maudslay Touch', *TNS*, 66 (1994-95), 153-74.
Exposition Universelle de 1855, *Rapports du Jury Mixte International,* I (1855).
Fairbairn, W., *On The Progress of Civil and Mechanical Engineering* (1859).
Farey, J., *Treatise on the Steam Engine* (1827).
Ferguson, E.S., *Early Engineering Reminiscences (1815-40), of George Escol Sellers* (1965).
Floud, R., *The British Machine Tool Industry, 1850-1914* (1976).
Forward, E.A., 'Simon Goodrich and His Work as an Engineer: Part 2', *TNS*, 18 (1937-38), 1-28.
Gilbert, K.R., *The Portsmouth Blockmaking Machinery* (1965).
Gilbert, K.R., *The Machine Tool Collection* (1966).
Gilbert, K.R., *Henry Maudslay* (1971).
Gilbert, K.R. 'Henry Maudslay 1771-1831', *TNS*, 44 (1971-2), 49-62.
Gilbert, K.R., *Early Machine Tools* (1975).
The Great Exhibition, *Official Catalogue,* I (1851).
Gregory, O., *A Treatise of Mechanics* (1806).
Griffiths, D., *Steam at Sea* (1997).
Habakkuk, H.J., *American and British Technology in the Nineteenth Century* (1962).
Hadfield, C., *Atmospheric Railways* (1985).
Hansard, T.C., *Typographia* (1825).

Hamilton, H., *An Economic History of Scotland in the Eighteenth Century* (1963).
Harris, T.R., *Arthur Woolf 1766-1837: The Cornish Engineer* (1966).
Henderson, W.O., *J.C. Fischer and His Diary of Industrial England, 1814-51* (1966).
Henderson, W.O., *Industrial Britain Under The Regency 1814-18: The Diaries of Escher, Bodmer, May and de Gallois* (1968).
Hesketh, E., *J.&E. Hall Ltd, 1785-1935* (1935).
Hills, R.L., *Power in the Industrial Revolution* (1970).
Hills, R.L., *Life and Inventions of Richard Roberts (1789-1864)* (2002).
Hogg, O.F.G., 'The Development of Engineering at the Royal Arsenal', *TNS*, 32 (1959-60), 29-42.
Holtzapffel, C., *Turning and Mechanical Manipulation*, II (1846).
Hopkins, T., 'Experiments and Observations on Diverging Streams of Compressed Air', *Manchester Literary and Philosophical Society, Memoirs*, 10 (1824-30), 9 March 1827.
Horner, J., *The Encyclopaedia of Practical Engineering and Allied Trades*, 10 vols (*c.*1910).
Hume, J.R. and Moss, M.S., *Beardmore – The History of a Scottish Industrial Giant* (1979).
Hyman, A., *Charles Babbage, Pioneer of the Computer* (1982).
International Exhibition of 1862, *Illustrated Catalogue of the International Exhibition: Catalogue of the Industrial Division, British Division*, I (1862).
International Exhibition of 1862, *Medals & Honourable Mentions awarded by the International Juries* (1863).
International Exhibition of 1862, *Reports by the Juries* (1863).
Jarvis, A., *Samuel Smiles and the Construction of Victorian Values* (1997).
Kargon, R.H., *Science in Victorian Manchester* (1977).
Kennison, G., *The Ancestors and Relatives of Robert Napier of West Shandon (1791-1876), Clyde Shipbuilder and Engineer*, with a forward by Major A.P.F. Napier (privately published, 2000).
Kilburn, T., *Joseph Whitworth, Toolmaker* (1987).
Knight, E.H., *The Practical Dictionary of Mechanics*, 3 vols (*c.*1870).
Landes, D.S., *The Unbound Prometheus* (1969).
Lea, F.C., *Sir Joseph Whitworth, A Pioneer of Mechanical Engineering* (1946).
Leigh, E., *Science of Modern Cotton Spinning* (1875).
Lineham, W.J., *A Textbook of Mechanical Engineering*, nineth edition, (1906).
Love, B., *The Hand-Book of Manchester* (1842).
Low, D.A. (ed.), *The Whitworth Book* (1926).
Lyon, D.J., *The Denny List* (1975).
Macleod, D., *Dumbarton, Vale of Leven, and Lochlomond: historical, legendary, industrial, and descriptive* (1884).
McDonald, K.C., *City of Dunedin: A Century of Civic Enterprise* (1965).
McDowell, D.M. and Jackson, J.D. (eds.), *Osborne Reynolds and Engineering Science Today* (1970).
McNeil, I., *Hydraulic Operation and Control Machines* (1954).
McNeil, I., *Joseph Bramah. A Century of Invention 1749-1851* (1968).
Meason, G., *The Official Illustrated Guide to the London and North-Western Railway* (*c.*1854).
Musson, A.E., 'Joseph Whitworth – Toolmaker and Manufacturer', *The Chartered Mechanical Engineer* (April 1963).
Musson, A.E., 'Joseph Whitworth – Toolmaker and Manufacturer', *Engineering Heritage*, 1 (1964).

Musson, A.E., 'The Life and Engineering Achievements of Sir Joseph Whitworth', *Joseph Whitworth 1803-1887: Exhibition to commemorate the centinary of his second term of Presidency of the Institution of Mechanical Engineers* (1966).

Musson, A.E. and Robinson, E., *Science and Technology in the Industrial Revolution* (1969).

Musson, A.E., 'The "Manchester School" and Exportation of Machinery', *Business History*, 14 (1972), 17-50.

Musson, A.E., 'Joseph Whitworth and the Growth of Mass-Production Engineering', *Business History*, 17 (1975), 109-49.

Musson A.E., *The Growth of British Industry* (1978).

Musson, A.E., 'The Engineering Industry' in Church, R.A. (ed.), *The Dynamics of Victorian Business* (1980).

Napier, D.D. and Bell, D. (eds.), *David Napier Engineer 1790-1869 – An Autobiographical Sketch with Notes* (1912).

Napier, J., *Life of Robert Napier of West Shandon* (1904).

Nicholson, T.R., *The Birth of the British Motor Car, 1769-1897* (1982).

Petree, J.F., 'The Lambeth Works of Maudslay, Sons and Field', *The Engineer*, 157 (1934), 585-86.

Petree, J.F., 'Maudslay, Sons & Field As General Engineers', *TNS*, 15 (1934-5), 21-36.

Petree, J.F., *Henry Maudslay, 1771-1831: and Maudslay, Sons & Field Ltd* (The Maudslay Society, 1949).

Petree, J.F., 'Henry Maudslay – Pioneer of Precision', *Engineering Heritage*, 1 (1964).

Pole, W. (ed.), *The Life of Sir William Fairbairn, Bart* (1877).

Porter, C.T., *Engineering Reminiscences* (reprinted 1985).

Pusely, D., *The Commercial Companion and Peerage of Commerce* (1860).

Rees, A., *Manufacturing Industry (1819-20)* (1972).

Roe, J.W., *English and American Tool Builders* (1916).

Rolt, L.T.C., *Isambard Kingdom Brunel* (1957).

Rolt, L.T.C. *Great Engineers* (1962).

Rolt, L.T.C., *Tools For The Job* (1965).

Rowlandson, T.S., *History of the Steam Hammer* (1864).

Rubinstein, W.D., *Capitalism, Culture and Decline in Britain 1750-1990* (1994).

Sandham, H., 'On the History of Paddle-Wheel Steam Navigation', *PICE* (1885), 121-59.

Saul, S.B., 'The Machine Tool Industry in Britain to 1914', *Business History*, 10 (1968), 22-43.

Saul, S.B. (ed.), *Technological Change: the United States and Britain in the Nineteenth Century* (1970).

Sennett, R. and Oram, H.J., *The Marine Steam Engine* (1918).

Scott, E.K., *Matthew Murray, Pioneer Engineer* (1928).

Sharp, R., 'A sad anniversary: the death of Maudslay, Sons & Field', *The Mariners Mirror*, 86 (2000), 75-78.

Slaven, A., *The Development of the West of Scotland 1750-1960* (1975).

Smiles, R., *Brief Memoir of the Late William Muir, Mechanical Engineer* (privately published, 1888).

Smiles, S., *Industrial Biography* (1863).

Smiles, S. (ed.), *James Nasmyth, Engineer: An Autobiography* (1883).

Smith, E.C., 'Joshua Field's Diary of a Tour through the Provinces, 1821, with Introduction and Notes', *TNS*, 13, (1932-3), 15-50.

Smith, E.C., *A Short History of Marine and Naval Engineering* (1937).

Sparey, L.H., *The Amateur's Lathe* (1972).
Spratt, H.P., *Outline History of Transatlantic Steam Navigation* (1950).
Spratt, H.P., *Handbook of the Collections Illustrating Marine Engineering* (1951).
Steeds, W., *A History of Machine Tools 1700-1910* (1969).
Sutton, J., *Lords of the East: The East India Company and its Ships* (1981).
Swade, D., *The Cogwheel Brain* (2000).
Tennent, Sir J.E., *The Story of the Guns* (1864).
Timbs, J., *The Year Book of Facts in the Great Exhibition of 1851* (1851).
Todd, W.B., *A Dictionary of Printers and other allied trades in London and vicinity* (1972).
Turnbull, J.G. (ed.), *A History of the Calico Printing Industry of Great Britain* (1951).
Turner, T., 'A History of Fenton, Murray & Wood', M.Sc. thesis, Manchester, UMIST, 1966.
Tyas, G.F., 'Matthew Murray. A Centenary Appreciation', *TNS*, 6 (1925-6), 111-43.
Vincent, W.T., *The Records of the Woolwich District* (1888-90).
Webb, S. and B., *The History of Trade Unionism* (1894).
Wiener, M.J., *English Culture and the Decline of the Industrial Spirit 1850-1980* (1981).
Whitworth, J., *On Plane Metallic Surfaces, and the Proper Mode of Preparing Them* (1840).
Whitworth, J., *Miscellaneous Papers on Mechanical Subjects* (1858).
Whitworth, J., *Papers on Mechanical Subjects* (1882).
Wilson, C.H. and Reader, W.J., *Men and Machines – A History of D.Napier & Son, Engineers, 1808-1958* (1958).
Woodbury, R. S., *History of the Lathe to 1850* (1961).
Woodcroft, B., *Brief Biographies of Inventors of Machines for the Manufacture of Textile Fabrics* (1863).
Woolrich, A.P., 'Joshua Gilpin's observations on paper-making in Britain and Ireland (1795-1796)', *The Quarterly* (Journal of the British Association of Paper Historians), Nos 20, 21 & 22, October 1996 and January and April 1997.
Woolrich, A.P., 'John Farey, jr., technical author and draughtsman: his contribution to Rees's *Cyclopaedia*', *Industrial Archaeology Review*, 20 (1998), 49-67.
Zagorskii, F.N. and Zagorskaia, I.M., *Henry Maudslay (1771-1831)* (1981).

CD ROM

A number of short papers relating to Henry Maudslay and Maudslay, Sons & Field were delivered to a seminar held at Kew Bridge Steam Museum on 26 July 2001. These papers together with images from the accompanying exhibition are published on a CD ROM. The papers are as follows:

Richard Maudslay	The Maudslay Family
Denis Smith	The Firm of Maudslay, Sons & Field
John Porter	The Life and Times of Kew's Maudslay Engine
John Cantrell	The Maudslay Legacy
Mary Mills	Shipbuilding at East Greenwich
Debbie Rudder	The Maudslay Beam Engine at Sydney's Powerhouse Museum
Michael Wright	Maudslay's Influence in Machine Design

Index

Amalgamated Society of Engineers, 116-119
American armaments industry, 120
Armstrong, William, 122-123
Atlas Works, 65-66

Babbage, Charles, 13, 16, 103-107, 137
Barlow, William Henry, 182
Barrel tail-stock, 100-101
Basset, Walter, 181
Belleville boiler, 178-181
Bentham, Samuel, 23, 107, 132, 166
Beyer, Charles Frederick, 64, 123
Beyer Peacock & Co., Manchester, 157
Birley family, 135
Boulton & Watt, Soho, 42-43, 166
Bourdon, François, 139-141
Braithwaite, John, 45, 47
Bramah & Robinson, Pimlico, 149
Bramah, Joseph, 18-19, 21-23, 26, 42, 94
Bridgewater Foundry, Patricroft, 135
Bright, John, 115, 122
Britannia Works, Manchester, 152-154
Brown, Charles, 182
Brunel, Isambard Kingdom, 15, 105, 121
Brunel, Marc, 18, 23, 105

Calico printing, 24, 74
Cardell, William Parminter, 89-90
Chadwick, Edwin, 111
Clement Driver, 100
Clement, Joseph, 13-16, 42, 48, 87, 94-108, 112
Coin-making manufacture, 25
Copland, Patrick, 94
Cotton, William, 87
Coulson, Jukes & Pelly, 45

Deverell, William, 141
Dickinson, Robert, 24
Donkin, Bryan, 44-46, 105-106

Drawing table, 97
Duncan, George, 175

Earl's Court Great Wheel, 180-181
Edinburgh Society of Arts, 129, 131
Edmonsen, Thomas, 149-151
Education of engineers, 116, 123-126
Enfield rifle, 121
Ewart, Peter, 87

Fairbairn, William, 18, 50-51, 53, 59, 107, 115
Farey, John, 23, 31-32
Field, Algernon Sydney, 178
Field, Joshua, 19, 20, 24, 28, 34, 46, 58, 105, 166-167, 172
Fischer, J. C., 48, 60
Fothergill, Benjamin, 68
Fox, James, 35, 51, 53

Galloway, Alexander, 47-50
Garnett family, 135, 153, 162-163
Gaskell, Holbrook, 135, 139, 141
Globe Works, 59, 61-62, 64, 68
Goddard, Ebenezer, 182
Goss, Henry, 148, 152
Great Exhibition, 1851, 67, 69, 88, 116, 137, 152
Gregory, Olinthus, 27, 31
Guages, 113

Hall, John, 44-6
Hampson, John, 111
Hayle Foundry, Cornwall, 47, 148
Herschel, William, 104
Hick, Benjamin, 37
Hill, James, 59, 62
Hill, Joseph, 79
Holtzapffel, Charles, 29-31, 44, 112, 149
Holtzapffel, John Jacob, 44, 112
Humphries, Francis, 139

Humphrys, George, 135

Imray, John, 182

James Watt & Co., Soho, 151, 157-158

Kirk, A.C., 182
Koechlin, Andre, 62-63

Lewis, Francis, 50, 55, 152
Locomotive manufacture, 64-66, 138
Lowry, Wilson, 95

Machinery
  Automaton, 87
  Block-making machinery, 23-24
  Bullet-making machine, 86, 89
  Difference engine, 102-106
  Double grindstone, 154, 160-161
  Drilling machine, 49, 156
  Drop stamp, 41
  Ellipse-drawing machine, 95-97
  Gear-cutting machine, 57
  Hydraulic press, 26, 42
  Jacquard punching machine, 69-71
  Key-grooving machine, 60, 136
  Lathe, 31-33, 42, 57, 98-101, 156
  Letter-copying press, 151, 158-160
  Lock-making machinery, 22-23, 41
  Measuring machine, 31, 113-114, 145
  Milling cutter, 103
  Nay-Peer, 81-83
  Nut-cutting machine, 132-133
  Perfecting machine, 80-83
  Pile driving machine, 42, 139
  Planing machine, 42, 57-58, 97-99
  Power loom, 59-60
  Punching & shearing machine, 25, 66-67, 69
  Radial drilling machine, 156
  Road-cleaning machine, 111-112, 149
  Self-acting spinning mule, 61-62, 68
  Shaping machine, 60, 136, 156
  Steam hammer, 134, 138-141
  Ventillating fan, 142
Machine tools, 39, 41-42, 51, 58, 71, 115-116, 125, 135-137, 154-158

Manchester Literary and Philosophical Society, 60-61
Manchester Mechanics Institution, 61
Manchester Steam Users Association, 115
Marine engine manufacture, 24-25, 71, 166, 168-182
Martineau, John, 40, 48
Martineau, Louis, 182
Maudslay, Henry (1771-1831), 12-16, 18-39, 49, 54-55, 66-67, 76, 84, 105, 111, 132-3
Maudslay, Henry (junior), 172, 175
Maudslay, John, 20
Maudslay, Joseph, 20, 168, 172
Maudslay, Thomas Henry, 20, 172
Maudslay, Walter Henry, 175, 177-178
Mass production, 15, 29, 54, 60, 126
May, J.G., 19
Mendham, John, 19-20
Morton, Thomas of Kilmarnock, 148, 164
Muir, Alfred, 153-154, 162-163
Muir, Andrew, 154, 162
Muir, Edwin, 154, 162
Muir, William, 13-16, 115, 147-165
Murray, Matthew, 35, 53

Napier, David (1788-1873), 13-16, 74-93
Napier, David (1790-1869), 76, 79, 81, 84-6
Napier, James Murdoch, 85, 87-89
Napier, John (1752-1813), 74, 76, 79
Napier, Robert (1726-1800), 74
Napier, Robert (1760-1847), 74-75
Napier, Robert (1791-1876), 76, 84
Nasmyth, Alexander, 130-132
Nasmyth, George, 51, 144
Nasmyth, James, 12-16, 26-28, 32-33, 111, 129-146, 149, 152

O'Connor, Charles, 182
Owens College, Manchester, 123

Penn, John, 45-46
Penn, Richard, 105
Pickering, John, 41
Plane surfaces, 28, 156-157

Registering compass, 87-89
Rennie, George, 42, 49
Rennie, John, 42, 46
Roberts, Richard, 12-16, 54-73
Rosse, 3rd Earl of, 87
Royal Small Arms Factory, Enfield, 161

Safety foundry ladle, 141-143
Sampson, John, 178-179
Schneider, Eugene, 139-141
Screw cutting, 29-31, 101-103
Seaward, John, 47-48
Seaward, Samuel, 27-28, 47-48
Sellers, George Escol, 31, 38
Sells, Charles, 174
Sennett, Richard, 182
Sharp, Robert Chapman, 59, 62
Sharp, Thomas, 59, 61-63, 66-68
Simpson, William, 75
Slide Rest, 26-28, 42
Smiles, Samuel, 12-13, 48-49, 141, 147, 153
Society of Civil Engineers, 41
Society of Millwrights, 46
Standard Screws, 28-31, 113
Steam engines, 137-139, 170-171
Steam Road Carriage, 63, 129-130
Stock selling, 142-144

Table engine, 24-26, 171-173
Taps and dies, 101-103
Taylor, Philip, 48
Telescopes, 87, 131
Tindale, Sarah, 19
Turret clocks, 67

Verbruggen, Jan, 41

Wallis, George, 120
Warriner, Henry, 174
Watt, James, 130, 141
Watt, James (junior), 166
Whitworth, Joseph, 12-16, 28, 62, 104, 106-107, 109-128, 142, 144, 149, 163
Whitworth rifle, 121-122
Whitworth scholarships, 123-124
Wilkinson, Joseph, 49, 106-107
Wilkinson, Thomas Jones, 54, 59, 62
Wilson, Robert, 141
Woodcroft, Bennet, 62, 148
Woolf, Arthur, 42, 45
Woolwich Arsenal, 18, 21, 26, 41-42, 161

Yeoman, Thomas, 41